This volume provides a brief but important summary of the essential tests that need to be performed on all new compounds, whether they be new drugs, pesticides or food additives, before they can be registered for use in the United Kingdom. These basic tests for mutagenicity were originally drawn up by the United Kingdom Environmental Mutagen Society (UKEMS) in 1983. They have now been fully revised by expert working groups from academia and industry under the auspices of UKEMS and in collaboration with the UK Department of Health. This volume therefore provides the latest official guidelines and recommendations for all scientists involved in the testing and registration of new compounds, not only in the UK, but in a wider international context.

The four main test procedures for measuring mutagenicity described in this volume are: bacterial mutation assays, metaphase chromosome aberration assays *in vitro*, gene mutation assays in cultured mammalian cells, and *in vivo* cytogenetics assays. Each of these tests is fully explained and described in practical and procedural detail, with additional information on the presentation and data-processing of results.

The volume will be essential for all toxicologists, pharmacologists and other scientists involved in regulatory affairs, mutagenicity testing and the successful registration of new chemical products.

Other titles of interest

Statistical Evaluation of Mutagenicity Test Data D. J. KIRKLAND
This companion UKEMS volume provides a rigorous and practical account of the design and statistical interpretation of mutagenicity tests.

A Practical Approach to Toxicological Investigations A. POOLE & G. B. LESLIE
This is a useful introduction to toxicology, largely in the context of testing new drugs. It addresses issues such as the selection and performance of suitable toxicological studies.

In-vitro *Methods in Toxicology* C. K. ATTERWILL & C. E. STEELE
This book describes the use of tissue-culture techniques to screen for drug-induced toxicity. It will be of great practical value to all professional toxicologists who are making use of these increasingly important techniques.

Basic mutagenicity tests

Basic mutagenicity tests: UKEMS recommended procedures

UKEMS sub-committee on guidelines for mutagenicity testing. Report. Part I revised

EDITOR

David J. Kirkland

ASSOCIATE EDITORS

David G. Gatehouse

David Scott

Jane Cole

Margaret Richold

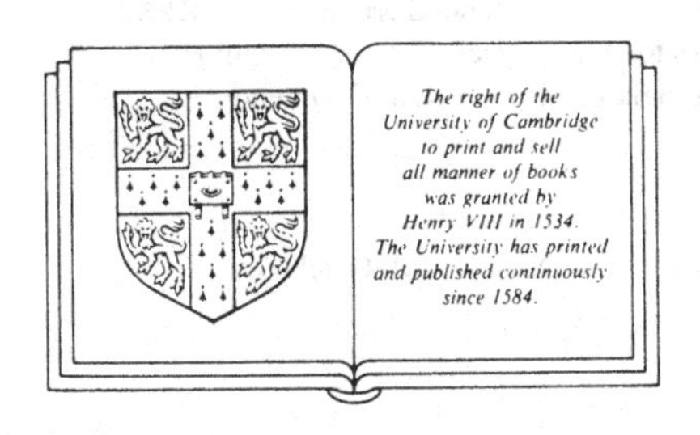

CAMBRIDGE UNIVERSITY PRESS

Cambridge

New York Port Chester

Melbourne Sydney

CAMBRIDGE UNIVERSITY PRESS
Cambridge, New York, Melbourne, Madrid, Cape Town, Singapore, São Paulo

Cambridge University Press
The Edinburgh Building, Cambridge CB2 2RU, UK

Published in the United States of America by Cambridge University Press, New York

www.cambridge.org
Information on this title: www.cambridge.org/9780521393478

First published 1990
This digitally printed first paperback version 2005

A catalogue record for this publication is available from the British Library

Library of Congress Cataloguing in Publication data
United Kingdom Environmental Mutagen Society. Sub-Committee on
Guidelines for Mutagenicity Testing.
Basic mutagenicity tests : UKEMS recommended procedures : UKEMS
Sub-Committee on Guidelines for Mutagenicity Testing : report /
editor, David J. Kirkland : associate editors, David G. Gatehouse
. . . [et al.]. – – Rev.
p. cm.
Includes bibliographical references.
1. Mutagenicity testing. I. Kirkland, David J. II. Gatehouse,
David G. III. Title.
QH465.A1U55 1990
616′.042– –dc20

90–34865
CIP

ISBN-13 978-0-521-39347-8 hardback
ISBN-10 0-521-39347-7 hardback

ISBN-13 978-0-521-01905-7 paperback
ISBN-10 0-521-01905-2 paperback

CONTENTS

STEERING GROUP

Margaret Fox, Paterson Institute for Cancer Research, Christie Hospital and Holt Radium Institute, Wilmslow Road, Manchester M20 9BX, UK

David J. Kirkland, Hazleton Microtest, University Road, Heslington, York YO1 5DU, UK

John Ashby, ICI plc, Central Toxicology Laboratory, Alderley Park, Macclesfield, Cheshire SK10 4TJ, UK

Bryn A. Bridges, MRC Cell Mutation Unit, Sussex University, Falmer, Brighton BN1 9RR, UK

Robert D. Combes, Inveresk Research International Ltd, Musselburgh EH21 7UB, UK

Brian J. Dean, Glaisdale, Main Street, Low Catton, Stamford Bridge, York YO4 1EA, UK

Robin J. Fielder, Department of Health, Hannibal House, Elephant and Castle, London SE1 6TE, UK

Susan A. Hubbard, Health and Safety Executive, Magdalen House, Stanley Precinct, Bootle, Merseyside L20 3QZ, UK (current address: ICI plc, Central Toxicology Laboratory, Alderley Park, Macclesfield, Cheshire SK10 4TJ, UK)

James A. Parry, School of Biological Sciences, University College of Swansea, Singleton Park, Swansea SA2 8PP, UK

CONTRIBUTORS

John Ashby, ICI plc, Central Toxicology Laboratory, Alderley Park, Macclesfield, Cheshire SK10 4TJ, UK

James Bootman, Life Science Research, Occold, Eye, Suffolk IP23 7PX, UK

Richard D. Callander, ICI plc, Central Toxicology Laboratory, Alderley Park, Macclesfield, Cheshire SK10 4TJ, UK

Ann Chandley, MRC Human Genetics Unit, Western General Hospital, Edinburgh EH4 2XU, UK

Jane Cole, MRC Cell Mutation Unit, University of Sussex, Falmer, Brighton BN1 9RR, UK

Natalie D. Danford, BIAS Ltd, The Innovation Centre, University of Swansea, Singleton Park, Swansea SA2 8PP, UK

Brian J. Dean, Glaisdale, Main Street, Low Catton, Stamford Bridge, York YO4 1EA, UK

Roy C. Forster, Italfarmaco, Via dei Lavoratori, 54, 20092 Cinisello Balsamo, Milano, Italy

Margaret Fox, Paterson Institute for Cancer Research, Christie Hospital and Holt Radium Institute, Wilmslow Road, Manchester M20 9BX, UK

R. Colin Garner, Cancer Research Unit, University of York, Heslington, York YO1 5DD, UK

David G. Gatehouse, Genetic and Reproductive Toxicology Department, Glaxo Group Research Ltd, Ware, Hertfordshire SG21 0DP, UK

Leigh Henderson, Huntingdon Research Centre, Huntingdon, Cambridgeshire, UK (Present address: Environmental Safety Laboratory, Unilever Research, Colworth House, Sharnbrook, Bedford MK44 1LQ, UK)

David J. Kirkland, Hazleton Microtest, University Road, Heslington, York YO1 5DU, UK

Douglas B. McGregor, Boehringer Ingelheim Pharmaceuticals Inc, 90 East Ridge, PO Box 368, Ridgefield, Connecticut 06877, USA (Present address: International Agency for Research on Cancer, 150 Cors Albert Thomas, 69372 Lyon Cedex 08, France)

Margaret Richold, Environmental Safety Laboratory, Unilever Research, Colworth House, Sharnbrook, Bedford MK44 1LQ, UK

Ian R. Rowland, British Industrial Biological Research Association, Woodmansterne Road, Carshalton, Surrey SM5 4DS, UK

David Scott, Paterson Institute for Cancer Research, Christie Hospital and Holt Radium Institute, Wilmslow Road, Manchester M20 9BX, UK

John Thacker, MRC Radiobiology Unit, Harwell, Didcot, Oxon OX11 0RD, UK

Philip Wilcox, Genetic and Reproductive Toxicology Department, Glaxo Group Research Ltd, Ware, Hertfordshire SG21 0DP, UK

LIST OF ABBREVIATIONS

2AAF= *N*-acetyl-2-aminofluorene
ACM= Advisory Committee on Mutagenesis (Canada)
APRT= adenine phosphoribosyl transferase
aprt= the gene for APRT
ASTM= American Society of Testing and Materials
8AZ= 8-azaguanine
B(a)P= benzo(a)pyrene
CA= chromosomal aberrations
CE= cloning efficiency
CHL= Chinese hamster lung (cells)
CHO= Chinese hamster ovary (cells)
COM= Committee on Mutagenicity (UK)
DHSS*= Department of Health and Social Security (UK)
DMN= dimethyl nitrosamine
DMSO= dimethyl sulphoxide
DOH*= Department of Health (UK)
EEC= European Economic Community
EMS= ethylmethane sulphonate
EPA= Environmental Protection Agency (USA)
FMN= flavin mononucleotide
GLP= Good Laboratory Practice
G6P= glucose-6-phosphate
his= histidine
his= gene for histidine
HPRT= hypoxanthine phosphoribosyl transferase
hprt= gene for HPRT
IARC= International Agency for Research on Cancer

* Name changed from DHSS to DOH between 1981 and 1989.

ICPEMC= International Commission for Protection against Environmental Mutagens and Carcinogens
IMF= induced mutant frequency
ip= intraperitoneal
ISCN= International System for Human Cytogenetic Nomenclature
JEPA= Japanese Environmental Protection Agency
JMHW= Japanese Ministry of Health and Welfare
JMITI= Japanese Ministry of International Trade and Industry
JMOL= Japanese Ministry of Labour
L5178Y= mouse lymphoma cell line
MFO= mixed function oxidases
MI= mitotic index
MN= micronucleus
MNNG= *N*-methyl-*N'*-nitro-*N*-nitrosoguanidine
MPE= micronucleated polychromatic erythrocyte
MRC= Medical Research Council (UK)
MTD= maximum tolerated dose
M I= first meiotic metaphase
M II= second meiotic metaphase
NADH= reduced nicotinamide adenine dinucleotide
NADP= nicotinamide adenine dinucleotide phosphate
NADPH= reduced NADP
NCE= normochromatic erythrocyte or normocyte
NCI= National Cancer Institute (USA)
4-NQO= 4-nitroquinoline-*N*-oxide
NTP= National Toxicology Programme (USA)
OECD= Organisation for Economic Co-operation and Development
OUA= ouabain
PCE= polychromatic erythrocyte
PHA= phytohaemagglutinin
SCE= sister chromatid exchange
SD= standard deviation
S9= post-mitochondrial supernatant
TFT= trifluorothymidine
6TG= 6-thioguanine
TK= thymidine kinase
TK6= human transformed lymphoblastoid cell line
TOSCA= Toxic Substances Control Act (USA)
TPA= 12-*O*-tetra-decanoyl phorbol ester
tryp= tryptophan

tryp= gene for tryptophan
UFAW= Universities Federation for Animal Welfare
UKEMS= United Kingdom Environmental Mutagen Society
V79= Chinese hamster fibroblast cell line

1

Introduction

D.J. KIRKLAND

1.1 GENERAL

1.1.1 Objectives

In March 1982 the United Kingdom Environmental Mutagen Society (UKEMS) appointed a Sub-committee to report on the minimal professional criteria that should be applied to mutagenicity testing in order to meet the requirements of the UK authorities. The tests recommended in the *Guidelines for the Testing of Chemicals for Mutagenicity* which was published by the Department of Health and Social Security (DHSS, 1981) formed the basis of the first report which dealt with the most commonly used mutagenicity tests (UKEMS, 1983). Other reports followed, dealing with *Supplementary Tests* (UKEMS, 1984) and *Statistical Evaluation of Mutagenicity Test Data* (UKEMS, 1989).

One objective of the Sub-committee was to ensure that these reports reflected the current state of knowledge in the field of mutagenesis, and it was suggested that each report be examined every 5 years to see if revisions were necessary. The first report was therefore due for examination in 1988.

Since the publication of the first report in 1983, many guidelines, recommendations and requirements for mutagenicity testing have been written. Those with the widest impact have been from the Organisation for Economic Co-operation and Development (OECD, 1983, 1984 and 1986), the European Economic Community (EEC, 1984), the Japanese Ministry of Health and Welfare (JMHW, 1984), the Japanese Ministry of Labour (JMOL, 1979), the Health Effects Guidelines of the US Environmental Protection Agency (EPA, 1982a, b), the Canadian Advisory Committee on Mutagenesis (ACM, 1986) and the American Society of Testing and Materials (ASTM, 1987). It was important, then, that any revision of the report on the basic test battery should take such guidelines into account.

As the expert Committee on Mutagenicity (COM) of the Department of Health was also revising its 1981 recommendations during 1988–9, (see Section 1.1.4) it was also most important that any revision of the UKEMS report on the basic test battery should reflect these changes. In fact, the changes have been quite significant in terms of both the numbers of tests considered in the basic test battery, and the protocol recommendations, but UKEMS and COM have liaised very closely during these revisions so that the respective recommendations enjoyed the support of the other party.

1.1.2 Terms of reference

The terms of reference of the Sub-committee at the time that Part 1 was prepared (UKEMS, 1983) were as follows:

1. to define the minimal criteria, i.e. the minimum basic experimental design required to perform current test procedures to professionally acceptable standards (with due regard to the Good Laboratory Practice Guidelines) of relevant authorities;
2. to define the criteria necessary to constitute a positive result in each of the current test procedures;
3. to define the criteria and extent of testing required to identify a material as non-mutagenic in each of the current test procedures;
4. to specify modifications to standard procedures which may be required to meet specific circumstances or to answer specific problems (e.g. procedures for testing volatile materials);
5. to prepare proposals for regular updating of the test systems and protocols as envisaged by the DHSS Guidelines and those of other bodies with regard to accepted technical advances supported by up-to-date and appropriate references;
6. to prepare recommendations for a framework of testing procedures subsequent to the initial battery, i.e. which supplementary test should be carried out and in what circumstances.

Terms (1) to (5) are still pertinent to the current revisions, although (6) is now superfluous in that the strategic element of the new COM recommendations addresses the framework of follow-up tests. Additionally, the extra emphasis on data generated *in vitro* means that, as for term (1), the protocol designs for these tests now need to be rigorous and robust rather than minimal.

1.1.3 Sub-committee structure

The Sub-committee consists of a Steering Group which is representative of industrial, academic, contract research and regulatory genetic toxicologists, and a series of Working Groups. The task of assessing and reporting on each of the procedures has been undertaken by the Working Groups which each consisted of a group leader plus four or five expert members. Only four Working Groups were established for this revision compared with six in 1983. This was a consequence of *Drosophila* and dominant lethal tests taking less prominent positions in the revised COM guidelines (DOH, 1989). The four Working Groups therefore addressed bacterial and mammalian cell mutation tests, and *in vitro* and *in vivo* cytogenetics assays.

1.1.4 The purpose of the revision

Apart from the fact that the field of genetic toxicology is still rapidly changing, and recommendations for testing therefore need to take account of advances in the science, the publication of many guidelines (see Section 1.1.1) and, in particular, the revision of the DOH recommendations, was a strong motivation to revise Part 1 (UKEMS, 1983). The fact that the basic package for screening purposes, according to DOH (1989), requires only two or three *in vitro* studies, demands that a very high standard of testing be applied to these studies. This has led to enlargement of some of the *in vitro* test protocol recommendations described herein, in particular the *in vitro* chromosomal aberration test which should now include an independent repeat experiment. It is also recommended that a more investigative approach be taken with all of the *in vitro* tests. In particular the second experiment of each *in vitro* study should include, as appropriate, emphasis on different concentrations of test chemical, different treatment or sample times, different S9 concentrations or even different treatment environments from those used in the first; an exact 'replicate' is not envisaged.

The new DOH recommendations (1989) also emphasise that the *in vivo* tests are primarily to investigate whether mutagenic potential seen *in vitro* can be expressed *in vivo*. These data are essential before any estimate of likely risk can be made. There is also a general desire to reduce use of animals as much as possible. The revised protocol recommendations for *in vivo* testing have managed to achieve both of these objectives. When used in this way there is no justification for further use of animals by repeating *in vivo* studies; this would be an unnecessary use of animals that could not be scientifically justified.

1.2 SPECIAL REQUIREMENTS

1.2.1 Safety

The safety of staff involved in the conduct of mutagenicity tests described in this and earlier reports has been a fundamental consideration of the Sub-committee. It has to be emphasised that staff should be fully trained in techniques for handling hazardous chemicals and potentially hazardous materials such as human blood. Containment areas for chemical handling and designated areas for handling potentially infected human cells should be delineated, and operating procedures should be in place to minimise handling hazards and to deal with spills. Disposal of waste containing mutagens/carcinogens or infected material also requires that pre-defined safety procedures be followed.

Details of handling precautions, disposal and spillage procedures are to be found in a number of texts for chemicals (IARC, 1979, 1980a; MRC, 1981; University of Birmingham, 1980) and in certain publications for pathogens and infectious material (e.g. Advisory Committee on Dangerous Pathogens, 1984).

1.2.2 Good Laboratory Practice

Most industrial, all contract and some academic laboratories will be conducting mutagenicity tests to provide reports to support regulatory submissions. Scientists in many of these laboratories will be well aware of the requirements of the Good Laboratory Practice (GLP) guidelines published by various organisations. However, not all countries operate GLP compliance programmes at this time, so, before conducting mutagenicity tests to support data submissions to the UK, USA, EEC and Japan in particular, it would be wise to become familiar with the following:

- Good Laboratory Practice. The UK Compliance Programme. DHSS, London, 1986.
- US Food and Drug Administration, Federal Register, 21 CFR Part 58; Good Laboratory Practice Regulations. December 22, 1978 (and its revisions of April 11, 1980 and September 4, 1987).
- US Environmental Protection Agency, Federal Register, 40 CFR Part 160; Pesticide Programmes; Good Laboratory Practice Standards. November 29, 1983.
- US Environmental Protection Agency, Federal Register, 40 CFR Part 792; Toxic Substances Control; Good Laboratory Practice Standards. November 29, 1983.
- OECD Principles of Good Laboratory Practice, C(81)30 (final) Annex 2. May 12, 1981.
- Japanese Good Laboratory Practice Guidelines of the Pharmaceutical Affairs Bureau, Ministry of Health and Welfare; Notification of March 31, 1982.

- Japanese Ministry of Agriculture, Forestry and Fisheries, GLP requirements on toxicological studies on pesticides; Notification 59 Nohsan No. 3850, issued by the Agricultural Production Bureau, August 10, 1984.

1.2.3 Technical expertise

Before embarking on any mutagenicity testing, or setting up a routine screening laboratory, it is essential that suitable experimental designs – 'base-line protocols' – be selected and documented. These enable consistent methods to be developed for every phase of each study, which are critical factors in successful testing. For this to occur, staff should be thoroughly trained in both theoretical and practical aspects of the methods they will use, if necessary by seeking (and taking) advice from investigators with proven experience in the field. Staff should be capable of carrying out every part of the test, and newly trained staff should be able to demonstrate their proficiency by correctly classifying the mutagenicity of a range of reference compounds.

1.3 TEST MATERIALS

1.3.1 Description

Testing substances 'blind' is to be strongly discouraged for routine testing. Details of the test material should be obtained from the supplier. It is important that the investigator obtains as much information as possible including, where appropriate:

- source
- batch number
- purity
- known impurities
- physical appearance
- chemical structure
- solubility
- reactivity in aqueous and non-aqueous solvents
- stability to temperature, pH and light
- storage conditions
- stability of solutions to temperature, pH and light
- adverse effects on man

If the full chemical name or CAS number can be obtained, and other relevant information such as colour index number for dyes, etc., this information should be reported. If the test material is a volatile liquid or gas, its boiling point and vapour pressure should be given. The investigator should check the appearance and characteristics of the test material on receipt against the supplier's description, and make an independent record of the physical appearance and conditions of storage.

1.3.2 Handling

Unless reliable data are available to the contrary, test substances should be handled as if they were toxic to man, unstable, and sensitive to light and heat. When testing substances which are destined for topical use, and where there is reason to suspect that the substances (e.g. sunscreens) may be photo-activated, either by design or incidentally, special controls need to be included or supplementary experiments should be performed to determine the effects of light of the appropriate wavelengths.

1.3.3 Preparation of solutions

Whenever possible, solutions of test substances should be freshly prepared immediately before each experiment and unused portions should not re-used unless there are appropriate stability data. When testing up to the limit of solubility in the test system, it may be necessary to examine a selection of prime solvents, and dilutions thereof, before deciding the best scheme. Maximum solubility would be defined as the highest concentration at which there is no visible precipitate. In all cases, however, the investigator must ensure that concentrations of organic solvents remain within tolerated levels, and have regard for the fact that use of an unusual prime solvent may necessitate inclusion of untreated controls, or even some preliminary investigations to ensure the solvent does not adversely affect spontaneous mutation/aberration frequencies or viability.

If positive results are obtained and a known impurity that gives rise to some concern (e.g. because of its structure) is present in the test substance, it is recommended that the impurity be assayed for mutagenicity at concentrations equivalent to those which would have been present in the positive concentration range of the parent test material used in the initial assay. If a mixture is to be tested, this should be stated. Since many mixtures are difficult to define chemically, details of their preparation should be provided. If it is known that one or more of the starting materials is a major constituent, it may be advisable to assay this in parallel with the mixture.

1.3.4 Special approaches

The comments above all apply to chemical test materials. If the test agent is physical in nature (e.g. radiation of various types) then a full description of the source, distance from source, time of exposure, means of exposure (i.e. with or without lid, or medium; static or roller bottles, etc.) and, preferably, a measure of the delivered 'dose' should be given.

The proposed sources and uses of the test substance should be known to the investigator since, for example, substances such as antibiotics, surfactants, preservatives, biocides and body fluids create special problems in bacterial mutation assays and may require special approaches.

1.4 POSITIVE CONTROLS

1.4.1 Choice of chemical(s)

The chemicals most widely used as positive controls are well documented in the literature, and often recommended in guidelines such as those of OECD (1983, 1984, 1986). The chemicals chosen should be those which the investigator has shown to act reproducibly in the system under test, and which identify satisfactory metabolic activation conditions, or discriminate between different bacterial tester strains, where appropriate. Different controls should therefore be used in non-activation and activation parts of any *in vitro* test.

It is often recommended that additional class-related positive controls should be used, for example specific azo-dyes and their amine reduction products, if an azo-reduction protocol is used (see Chapter 2), or hydrazines when hydrazine-containing drugs are under test, as recommended by Ashby & Purchase (1979). Such procedures are to be encouraged where appropriate, but it is recognised that, with many compounds, it may not be possible to identify a class-related mutagen.

1.4.2 Choice of concentrations

It is recommended that prior to a testing programme, each laboratory should construct dose-response curves for each positive control mutagen. The test concentration chosen for each reference mutagen should then be set at a level which is close to the limits of detection and which would be expected to test the performance and resolution of the assay, rather than at a level which will always give more than adequate responses irrespective of the sensitivity of the assay on any given occasion. An alternative, and equally acceptable, approach to utilising a low dose of a potent mutagen, is to utilise a moderate dose of a weak mutagen. Either way, the investigator must be able to demonstrate that the test system is able to consistently detect small increases in mutant/aberration frequency.

1.5 EXOGENOUS METABOLIC ACTIVATION

1.5.1 The use of metabolising systems

Metabolising systems that 'optimise' the activation systems present in mammals need to be provided in *in vitro* screening tests to facilitate the detection of as wide a range of chemical classes as possible. When used in basic research, the priorities may be different: for example, strenuous efforts might be made to optimise the generation of reactive metabolites from refractory chemicals, or to investigate the importance of different metabolic pathways. For screening purposes it is more important that the metabolic systems adequately detect a *wide* range of chemicals, rather than allow analysis in depth of a *narrow* range of chemicals. The most commonly used system is the post-mitochondrial supernatant (S9) from the livers of rats pre-treated with an enzyme inducer such as Aroclor 1254, and this has been found to be a useful system for general screening of chemicals. Further discussion is given in the individual chapters.

1.5.2 Preparation and characterisation of S9

The preparation of S9 is a relatively simple operation, requiring, however, close attention to detail (Ames *et al.*, 1975; Maron & Ames, 1983). All apparatus and solutions should be sterile, and the preparation should be kept at or near 0°C at every stage in the process. The tissue should be homogenised in a manner allowing reproducibility from batch to batch. The Ultra-turrax or Potter homogenisers give satisfactory results. A useful guide to consistency is the protein concentration of the S9, which can be determined very simply by standard colorimetric methods, and which should be used as a bench-mark against which successive batches of S9 may be referred. There should be no necessity for filtering S9 preparations to ensure sterility, providing that clean animals are used and good aseptic technique is employed. Detergent/disinfectant cleaning of animals can considerably reduce subsequent contamination. It is not advisable to try to 'clean up' heavily contaminated preparations, and they should be discarded. S9-mix should also be discarded at the end of the experimental period (i.e. storage overnight is not acceptable).

Ready-made S9 preparations can now be purchased: there is no intrinsic objection to their use, provided that their source and method of preparation is well documented in the report. Quality control details from the supplier should also be documented. Whatever the source of the S9, it is imperative that its activity be checked, using reference mutagens (see Chapter 2), prior to or during any experiment where metabolic activation is used. It should be noted that S9 as described above contains both microsomes and cytosol. These are different fractions that may metabolise some chemicals differently, and, although there is no cause to use the separate fractions in routine screening, they may be useful in follow-up studies. Their methods of preparation are different (see Pyykko, 1983).

1.5.3 Storage of S9

Some investigators advocate the use of fresh S9 preparations for each experiment. Clearly this is impracticable in laboratories screening large numbers of compounds. Extensive studies of the effect of storage on activating capacity of S9 towards a wide range of mutagens have not been reported. However, there are numerous, more limited, investigations particularly of the activity of mixed-function oxidase enzymes in stored S9. Although there is considerable variation in the findings (summarised by Hubbard *et al.*, 1985), the following general points emerge. Storage of S9 at −20 °C is unsatisfactory due to rapid loss (in 3–5 days) of enzyme activities (Litterst *et al.*, 1974). Temperatures of −70 °C or −196 °C (liquid nitrogen) provide more stable storage conditions. Usually there is an initial 10–20% loss of enzyme activity (Gatehouse, 1987), probably due to freezing and thawing. Thereafter there is a much more gradual decline, although the rate is dependent on the particular enzyme being measured (Ashwood-Smith, 1980; Hubbard *et al.*, 1985) with up to 50% of the original enzyme activity being lost after 3 months in some cases. There is general agreement that periods of storage greater than 1 year are unsatisfactory and both IARC (1980b) and JMHW (1984) recommend that S9 should not be stored for longer than 3 months. As in other aspects of short-term testing, somewhat arbitrary decisions must be made, but a consistent method of working should be adopted and suitable positive controls requiring activation always be included.

1.5.4 S9-mix

Discussion on the effects of using different concentrations of S9 in co-factors (referred to as S9-mix) is given in the individual chapters.

1.5.5 Hepatocytes

The use of intact hepatocytes for activation has been described for a number of mammalian cell and microbial systems, and hepatocytes from uninduced and Aroclor 1254-induced animals have been used. Some reports have suggested that hepatocyte-mediated assays are relatively insensitive in detecting known indirect mutagens, and modifications to the induction process (e.g. 1 day instead of 5 days pre-treatment) have been suggested as improvements. These matters are discussed by Hass *et al.* (1985).

1.6 PRESENTATION OF RESULTS

1.6.1 Minimum data to be presented

The description of the experimental design should be detailed enough to allow independent replication of the assay. Precise details of the protocol should be provided: if a published protocol was used, this should be referred to, and any deviations from it should be indicated. The source and the method of preparation of the exogenous metabolic activation system should be given, together with details of any inducers which were used. If an S9-mix was used, the percentage of S9 in the S9-mix and the concentration of buffers and co-factors should be given. Also it is important to record the final concentration of S9 in the test system. Any items bought in from proprietary sources (e.g. S9, ready-poured plates, media) should be noted with details of sources and the above information.

For regulatory purposes, 'raw' data should be provided, including results obtained for negative and positive controls. Such data should always be given in addition to mutation frequencies or other transformations. Presentation of raw data allows independent statistical evaluation.

1.6.2 Data processing

Data analysis methods should be carefully documented or referenced. UKEMS working groups have published their own recommendations for analysis of data obtained from the assays discussed herein (UKEMS, 1989).

1.6.3 Publication of data

Publication of data in scientific journals is strongly encouraged wherever possible. Not only does this ensure that robust protocols are used to generate high quality data, but it enables others to evaluate the performance of various protocols, which is critical in the development of testing strategies.

1.7 REFERENCES

ACM (1986). *Guidelines on the Use of Mutagenicity Tests in the Toxicological Evaluation of Chemicals*. A Report of the DNH & W/DOE Environmental Contaminants Advisory Committee on Mutagenesis. National Health and Welfare and Environment, Canada.

Advisory Committee on Dangerous Pathogens (1984). *Categorisation of Pathogens according to Hazard and Categories of Containment* (ISBN 011 883761 3).

Ames, B.N., McCann, J. & Yamasaki, E. (1975). Methods for detecting carcinogens and mutagens with the *Salmonella* /mammalian microsome mutagenicity test. *Mutation Research*, **31**, 347–64.

Ashby, J. & Purchase, I.F.H. (1979). The selection of appropriate chemical class controls for use with short-term tests for potential carcinogenicity. *Annals of Occupational Hygiene*, **20**, 297–301.

Ashwood-Smith, M. J. (1980). Stability of frozen microsome preparations for use in the Ames *Salmonella* mutagenicity assay. *Mutation Research*, **69**, 199–200.

ASTM (1987). Guidelines for Minimal Criteria of Acceptability for Selected Short-term Assays for Genotoxicity, ed. R. W. Naismith. *Mutation Research*, **189**, 81–183.

DHSS (1981). Guidelines for the Testing of Chemicals for Mutagenicity. Prepared by the Committee on Mutagenicity of Chemicals in Food, Consumer Products and the Environment, Department of Health and Social Secuiity. *Report on Health and Social Subjects, No. 24*. Her Majesty's Stationery Office, London.

DOH (1989). *ibid*. Department of Health. *Report on Health and Social Subjects No. 35*.

EEC (1984). Methods for the determination of physico-chemical properties, toxicity and ecotoxicity; Annex V to Directive 79/831/EEC. *Official Journal of the European Communities No. L251*, 131–45.

EPA (1982a). *Health Effects Test Guidelines*. Office of Toxic Substances, Environmental Protection Agency, Washington, USA, August 1982.

EPA (1982b). *Pesticide Assessment Guidelines, Subdivision F. Hazard Evaluation: Human and Domestic Animals*. Office of Pesticides and Toxic Substances, Environmental Protection Agency, Washington, USA, October 1982.

Gatehouse, D. (1987). Guidelines for testing of environmental agents. Critical features of bacterial mutation assays. *Mutagenesis*, **2**, 397–409.

Hass, B.S., Heflich, R.H., Shaddock, J.G. & Casciano, D.A. (1985). Comparison of mutagenicities in a *Salmonella* reversion assay mediated by uninduced hepatocytes and hepatocytes from rats pretreated for 1 or 5 days with Aroclor 1254. *Environmental Mutagenesis*, **7**, 391–403.

Hubbard, S.A., Brooks, T.M., Gonzalez, L.P. & Bridges, J.W. (1985). Preparation and characterisation of S9 fractions. In *Comparative Genetic Toxicology*, ed. J.M. Parry and C.F. Arlett. MacMillan Press, Basingstoke, pp. 413–38.

IARC (1979). Some halogenated hydrocarbons. *IARC Monographs on the*

Evaluation of the Carcinogenic Risk of Chemicals to Humans, 20. International Agency for Research on Cancer, Lyon, pp. 85–106.

IARC (1980a). *Laboratory Decontamination and Destruction of Aflatoxins B', B_2, G_1, G^2 in Laboratory Wastes*, ed. M. Castegnaro, D.C. Hunt, E.B. Sansone, P.L. Schuller, M.G. Siriwardana, G.M. Telling, H.P. von Egmond and E.A. Walker. IARC Scientific Publications. International Agency for Research on Cancer, Lyon.

IARC (1980b). Report 10. Basic requirements for *in vitro* metabolic activation systems in mutagenesis testing. In *Long-term and Short-term Screening Assays for Carcinogens: a Critical Appraisal, IARC Monographs on the Evaluation of the Carcinogenic Risk of Chemicals to Humans, Supplement 2*. International Agency for Research on Cancer, Lyon, pp. 277–94.

JMHW (1984). *Information on the Guidelines of Toxicity Studies Required for Applications for Approval to Manufacture (Import) Drugs (Part 1)*, Notification No. 118 of the Pharmaceutical Affairs Bureau. Ministry of Health and Welfare, Japan.

JMOL (1979). *On the Standards of the Mutagenicity Test Using Micro-Organisms, The Labour Safety and Hygiene Law*, The Labour Standard Bureau, Ministry of Labour, Japan.

Litterst, C.L., Mimnauch, E.G., Reagan, R.L. & Gran, T.E. (1974). Effect of storage on microsomal mixed function oxidase activity in the mouse liver. *Biochemical Pharmacology*, **23**, 2391–4.

Maron, D.M. & Ames, B.N. (1983). Revised methods for the *Salmonella* mutagenicity test. *Mutation Research*, **113**, 173–215.

MRC (1981). *Guidelines for Work with Chemical Carcinogens in Medical Research Council Establishments*. Medical Research Council, London.

OECD (1983). *OECD Guideline for Testing of Chemicals. Genetic Toxicology, No. 471–474*. Organisation for Economic Cooperation and Development, Paris, 26 May 1983.

OECD (1984). *ibid, No. 475–478*. 4 April 1984.

OECD (1986). *ibid, No. 479–485*. 23 October 1986.

Pyykko, K. (1983). Characterisation and stability of rat liver microsomes isolated by a rapid gel filtration method. *Acta Pharmocologica et Toxicologica*, **52**, 39–46.

UKEMS (1983). *UKEMS Sub-committee on Guidelines for Mutagenicity Testing. Report. Part I. Basic Test Battery*, Ed. B.J. Dean. United Kingdom Environmental Mutagen Society, Swansea.

UKEMS (1984). *ibid, Part II, Supplementary Tests*.

UKEMS (1989). *ibid. Part III, Statistical Evaluation of Mutagenicity Test Data*, Ed. D. J. Kirkland. Cambridge University Press.

University of Birmingham (1980). *Rules and Notes of Guidance for the Use of Chemical Carcinogens in the University*. The University of Birmingham, Birmingham, UK.

2

Bacterial mutation assays

D.G. GATEHOUSE
P. WILCOX
R. FORSTER
I.R. ROWLAND
R.D. CALLANDER

2.1 INTRODUCTION

2.1.1 Principles and genetic basis of the assay

The use of bacteria for screening chemicals for their potential mutagenicity and carcinogenicity to human beings is based on the observation that the primary structure of the genetic material, DNA, is the same throughout the living world. Short-term tests detecting bacterial mutation measure the ability of physical or chemical agents to damage DNA. There is ample evidence that DNA damage in the germ cells of metazoans (e.g. insects, vertebrates) can cause heritable genetic defects. A considerable body of data supports the idea that DNA damage in somatic cells may be a critical event in the initiation of cancer. Since DNA is susceptible to damage at numerous sites and by a variety of different mechanisms, it is technically difficult to measure the damage directly in all its manifestations. On the other hand, it is relatively straightforward to use the induction of mutation in bacteria as a very sensitive *indirect* indicator of DNA damage. Bacteria can be grown in large numbers overnight, enabling the detection of very rare mutational events. Extensive knowledge of bacterial genetics has allowed the construction of special strains of bacteria which are much more sensitive than are wild-type strains to a variety of agents.

Mutation may be defined as a stable heritable (replicable) change in a DNA nucleotide sequence: such heritable changes may be due to base-substitutions (transitions, transversions); frameshifts (deletions or additions of one or a few nucleotide pairs leading to a change in the reading frame of the genetic code); large deletions; insertions or translocations. Few if any mutagens induce only one type of mutational change. Rather, most mutagenic agents exhibit a characteristic mutagenic spectrum which depends on several factors, including the nature of the primary DNA alterations (e.g. modifications of bases, phosphate or sugar residues, strand breaks, or incorporation of modified bases), and the subsequent secondary effects caused by the response of the organism to these DNA modifications. These secondary effects might include the operation of various forms of DNA repair, and the replication of daughter strands on modified templates. The same mutagen may therefore produce different mutagenic spectra in organisms with different genetic backgrounds.

Mutagenic effects may be scored by methods which detect '**forward**' or '**reverse**' mutation. Genetic systems which detect forward mutations have the theoretical advantage of presenting a large target for the mutagen to attack and should therefore enable the detection of many of the changes mentioned above (including deletions), all of which may be expressed in the same phenotype. However, a forward-mutation system may fail to detect an increase in the induced frequency of a rare event against a background of common events. Acquisition of resistance to a toxic chemical (e.g. an amino-acid analogue or drug) is a frequently used genetic marker for forward mutation.

Reversion assays employ bacteria which are **already mutant** at a locus whose phenotypic effects are easily detected: the assay determines the frequency at which exposure to the test-chemical abolishes or suppresses the effect of the pre-existing mutation. The genetic target presented to the test-chemical is therefore small, specific and selective. Several bacterial strains, or a single strain with multiple markers, are necessary to overcome the effects of mutagen specificity. The reversion of bacteria from growth-dependence on a particular amino acid to growth in the absence of that amino acid (reversion from auxotrophy to prototrophy) is the most widely used marker in reverse-mutation assays.

In order to make bacteria more sensitive to mutation by chemical and physical agents, several additional traits have been introduced: these include various DNA-repair deficiencies, increased permeability of the bacterial cell wall to bulky hydrophobic chemicals, and the introduction of plasmids which confer on the bacteria increased sensitivity to mutation without a concomitant increase in sensitivity to the lethal effects of test compounds. In some cases plasmids have been shown to confer resistance to the lethal effects of certain compounds (Nunoshiba & Nishioka, 1984; Thomas & Cole, 1986).

In some systems, notably the *Salmonella* test, further sensitivity is gained by the fact that the initial mutation responsible for the auxotrophic growth requirement is situated at a site within the gene (hisG, hisC or hisD) that is particularly sensitive to reversion by specific classes of mutagen, i.e. the particular sites are 'hot spots' for mutation. With one of the more recently developed strains (TA102) the hisG gene, containing an ochre mutation at hisG428, is located on a multicopy plasmid, pAQ1 (Levin *et al.*, 1982a). As each bacterial cell contains approximately 30 copies of the plasmid, there are multiple copies of the mutated gene within each cell and reversion of any of these will result in a his^+ phenotype.

For most of the commonly used strains the DNA sequence around the original histidine mutation has been determined (Table 2.1). DNA sequence analysis of spontaneous and chemically induced his^+ revertants has revealed that multiple modes of reversion to a histidine-independent phenotype are possible in each instance (Barnes *et al.*, 1982; Hartman *et al.*, 1986; Hartman & Aukerman, 1986). For example, revertants of hisG428 (the mutation in TA102 and TA104) included true revertants to the exact wild type sequence (AT→GC transitions), intragenic suppressors near the initial mutation (AT→TA and AT→GC), and three base-pair or six base-pair deletions (Levin *et al.*, 1984). Also, because the original mutation results in an ochre nonsense codon, extragenic suppressors could occur at four different tRNA structural genes (these included GC→AT transitions, and CG→AT and CG→GC transversions). Thus the hisG428 mutation can be reverted by all six possible base-pair changes, as well as small deletions (Hartman *et al.*, 1986; Hofnung & Quillardet, 1986). Therefore, even though a mutagen may be specific in reverting only one of the battery of tester strains, it is not possible to deduce the exact nature of the mutagenic event on a molecular level. There may be instances where it would be desirable to obtain such information and for this eventuality additional strains are being developed that can only be reverted by specific base-pair substitution mutations (Levin & Ames, 1986).

The advantages of using bacteria to detect mutagenic activity are evident from the foregoing paragraphs. However, the bacteria most commonly used in these assays (*Salmonella typhimurium* and *Escherichia coli*) do not possess the enzyme systems which, in mammals, are known to mediate the transformation of many classes of chemically unreactive, but nevertheless mutagenic and carcinogenic compounds, to highly reactive electrophiles which can form adducts with nucleophilic sites in DNA (Miller & Miller, 1976; Sims, 1980). In order to overcome this major drawback, means have been devised to supply bacteria with the required enzymes in the form of cell-free extracts prepared from laboratory rodents. The most commonly used preparation consists of a post-mito-

Table 2.1. *Histidine mutations in commonly used* Salmonella *strains*

Mutation	Strain	Nature of mutation	Reversion events
hisG428	TA102 TA104	w/t CAG–AGC–AAG–CAA–GAG–CTG– ↓ mutant CAG–AGC–AAG–TAA (ochre)	All possible transitions and transversions Extragenic suppressors Small deletions (−3, −6)
hisG46	TA100 TA1535	w/t GTG–GTC–GA*T–CTC–GGT–ATT– ↓ mutant –CCC–	Subset of base-pair substitution events Extragenic suppressors
hisD6610	TA97	w/t GTC–ACC–CCT–GAA–GAG–A*TC–GCC mutant GTC–ACA–CCC–CCC–TGA (opal)	Frameshifts
hisD3052	TA98 TA1538	w/t GAC–ACC–GCC–CGG–CAG–GCC–CTG–AGC mutant GAC–ACC–GCC–GGC–AGG–CCC–TGA (opal)	Frameshifts
hisC3076	TA1537	w/t sequence not known mutant presumed +1 near CCC	Frameshifts

A*, 6-methyldeoxyadenosine.
w/t, wild-type.

chondrial supernatant ('S9') of homogenised liver tissue from rats. This material contains a variety of drug-metabolising enzymes, including cytochrome-dependent monooxygenases, cytochrome-independent oxidases, amidases, esterases, glutathione S-transferases, sulphotransferases, acyl and methyl transferases, dehydrogenases, esterases and peroxidases.

The S9 is supplemented with co-factors, salts and a buffer system, the complete preparation being known as the 'S9-mix'. Methods for preparation and use of S9-fraction are given in more detail in Section 2.3.2.5.

2.1.2 Relevance and limitations

Bacterial mutation tests have been subject to several large-scale trials designed to test both their relevance and reliability in detecting mutagens (Ames *et al.*, 1975; McCann & Ames, 1976; Purchase *et al.*, 1978; McMahon *et al.*, 1979; Rinkus & Legator, 1979; IARC, 1980a; Bartsch *et al.*, 1980; De Serres & Ashby, 1981; Venitt, 1982; Mortelmans *et al.*, 1985; Zeiger, 1987; Tennant *et al.*, 1987). These studies were primarily concerned with assessing the correlation between results obtained in bacterial assays and the carcinogenic activity of chemicals. The theoretical basis for such a correlation has been strengthened considerably over recent years by the demonstration that the induction of point mutations at specific sites within proto-oncogenes is one of the mechanisms of oncogene activation (Tabin *et al.*, 1982; Reddy *et al.*, 1982; Capon *et al.*, 1983; Fasano *et al.*, 1984).

Most of the studies cited above suggest that there is a good **qualitative** relationship (around 80–90% sensitivity and specificity) between mutagenicity in the *Salmonella* assay and carcinogenicity for many, although not all, chemical classes. However, the two most recent papers (Tennant *et al.*, 1987; Zeiger, 1987) have reported much lower sensitivity values for the *Salmonella* assay (45% and 54%, respectively) and this has reopened questions concerning the value of this assay for predicting carcinogenic activity. Both of these papers are based on the results obtained with chemicals that have been tested for carcinogenicity in the rat and mouse by the National Cancer Institute (NCI) or the National Toxicology Programme (NTP). Ashby & Tennant (1988) carried out a re-evaluation of some of these chemicals which involved classifying the carcinogens into various sub-classes based on the reproducibility of response between species and sex, and whether tumours were induced at more than one site. When they considered those carcinogens that were active in both rats and mice or agents found to be carcinogenic to only a single species but active at two or more sites in that species, the sensitivity of the *Salmonella* assay increased to 70%. This indicates that the low sensitivities reported by Zeiger (1987) and Tennant *et al.* (1987) were due largely to the poor predictivity of the *Salmonella* assay for carcinogens that were tumorigenic at only a single site in one species (classes C and D in Ashby & Tennant, 1988). It is possible that many compounds that fall into these catagories may be carcinogenic by 'non-genotoxic' mechanisms and, if this is the case, it is unreasonable to expect the *Salmonella* assay to predict this activity.

The relevance of such tests to the detection of mutagens is self-evident, but the most commonly used reverse-mutation tests cannot detect large deletions, nor those heritable genetic changes characteristic of higher organisms whose DNA is organised into chromosomes. Such changes include certain structural chromosome damage and numerical changes to the karyotype. The formation of the latter is to a great extent dependent on interference with the complex spindle apparatus which segregates chromosome-sets between daughter cells during mitosis and meiosis in diploid eukaryotes. Thus, bacterial mutation tests are not regarded as reliable indicators of heritable genetic risk to the germ cells (ICPEMC, 1983; Bridges & Mendelsohn, 1986).

An interesting new development in the field of environmental mutagenesis is the suggestion that mutations in somatic cells may be responsible for various pathological conditions in man, other than cancer. Possible examples include atherosclerotic plaques, senile cataracts, gallbladder disease and certain gastro-intestinal metaplasias and ulcers (Hartman, 1983). Somatic mutations may also be important in the general ageing process in higher organisms (Hartman & Aukerman, 1986). Thus, the impact of somatic mutations on human health may be much more widespread than the current emphasis on the mutagen/cancer relationship (Hartman, 1983).

The need to supply bacteria with an exogenous metabolising system, in order to mimic the well developed systems of intact mammals, is an important limitation to the extrapolation of results gained in bacterial tests to judge the potential mutagenicity and carcinogenicity of chemicals to man. Rapidly dividing, DNA-repair deficient bacteria treated *in vitro* with the test substance in the presence of a crude S9 preparation are not subject to the checks and balances afforded by the mammalian patterns of absorption, distribution, metabolism (activation, detoxification) and excretion which can greatly modify the response of a particular species and strain of mammal to the mutagenicity or carcinogenicity of a given compound. In the analysis of Ashby & Tennant (1988) about

30% of compounds which were positive in the *Salmonella* assay were not carcinogenic in the mouse or rat. Many of these compounds were direct-acting mutagens and their lack of carcinogenic activity is probably due to non-absorption or preferential detoxification *in vivo*. Such limitations must be taken into account when assessing the relevance of bacterial mutation tests. A positive result in such a test cannot directly classify an agent as mutagenic or carcinogenic to man, or other mammals. Rather, a positive bacterial mutation test should be regarded as an early warning of potential hazard and an indication that further work in higher organisms is required.

Another limitation of bacterial mutation assays is that they may be unsuitable for testing compounds which are very toxic to the bacterial tester strains, such as antibiotics which are effective against Gram-negative strains. Negative results at low μg/plate concentrations are of little value and in such cases another type of test system should be considered.

Despite these limitations, short-term tests employing bacterial mutation provide a sensitive means for detecting physical and chemical agents which can interact or modify the function of DNA and, as such, must remain the method of choice for primary genotoxicological screening. Bacterial mutation tests in general, and the *Salmonella* assay in particular, have been studied, validated, characterised and understood in far greater detail than most other genotoxicological or toxicological assays.

2.2 THE TEST MATERIAL

2.2.1 Properties of the test material

General points for consideration when testing chemical entities have been discussed in Chapter 1.

If the test substance is of biological origin (e.g. a foodstuff or biological fluid) the content of free amino acids, especially those used as markers in reverse-mutation tests (i.e. histidine/tryptophan), should be known. Several laboratories have reported on the problems of testing urine and faecal concentrates due to the presence of significant quantities of these amino acids (Gibson *et al.*, 1983; Venitt & Bosworth, 1983). The testing of food concentrates for mutagenicity is also complicated by the presence of growth-stimulating or inhibitory substances (Aeschbacher, 1980). The presence of histidine in the test material can cause statistically significant increases in the number of revertants obtained, especially with bacterial strains measuring base-substitution mutations (Aeschbacher *et al.*, 1983). In addition, the presence of any chemicals which can be utilised as 'surrogate intermediates' in the tryptophan or histidine biosynthetic pathways, at points after the mutational block, will lead to artificial increases in reversion rates. Novel chemical entities containing indole or imidazole rings can be particularly troublesome in this respect (D.G. Gatehouse, personal communication).

For the *E. coli* WP2 series of test strains, the mutation conferring tryptophan dependence is at the *tryp E* locus, and blocks the conversion of chorismic acid to anthranilic acid. Thus the presence of any chemicals which can be utilised as intermediates in the biosynthetic chain beyond chorismic acid will enable the auxotrophic *E. coli* cells to synthesise tryptophan and grow, even though no mutational event has occurred within these cells. Similarly, the mutation at the hisG locus which blocks histidine biosynthesis in *S. typhimurium* strains TA1535 and TA100 allows the utilisation of chemical intermediates in the biosynthetic chain after phosphoribosyl pyrophosphate. Treat-and-plate tests can be used to resolve these problems (Combes *et al.*, 1984; Rowland *et al.*, 1984).

Treatment of the samples prior to assay with XAD-2 resin, or with enzymes such as histidine decarboxylase, may be of limited value. However, it would seem sensible to utilise alternative assay procedures to those measuring reversion to prototrophy, for the evaluation of such complex biological fluids. The use of forward mutation assays to 8-azaguanine resistance (Skopek *et al.*, 1978); ampicillin resistance (Solt & Neale, 1979; Gatehouse & Paes, 1983); L-arabinose resistance (Pueyo & Hera, 1986) or rifampicin resistance (Vithayathil *et al.*, 1983) might provide more definitive data.

2.2.2 Selection of solvents

The solvent should not be toxic to the indicator bacteria at the concentration to be used in the assay, and must be chosen to take account of the physical and chemical properties of the test substance. Where possible the choice of solvent should be made in consultation with the chemist who prepared the compound taking into account the stability of the material in the chosen solvent. The nature and concentration of the solvent used may have a marked effect upon the test result. It is still common practice to use dimethyl sulphoxide (DMSO) as the solvent of choice for hydrophobic compounds, since it is non-toxic to bacteria up to the concentrations normally used, and it is freely miscible with water. However, there are a number of disadvantages to the use of this solvent. DMSO decomposes/oxidises with time on exposure to air to form toxic and mutagenic products. The latter induce base-substitution mutations in bacterial cells. It is therefore essential to use batches of the highest purity grade available.

DMSO is not an inert chemical and it may react with the test material resulting in either an enhancement or a reduction of the observed mutagenic effect. Such phenomena may be qualitative in nature. As far back as 1977, DMSO was shown to inhibit the mutagenicity of dimethylnitrosamine and diethylnitrosamine (Yahagi *et al.*, 1977). Recently, there have been numerous examples illustrating the way in which DMSO can enhance rather than inhibit mutagenesis, e.g. p-phenylenediamine (Burnett *et al.*, 1982), 2-aminoanthracene (Anderson & McGregor, 1980) and hexachloroacetone (Nestmann *et al.*, 1985).

Furthermore, DMSO and other organic solvents (methanol, acetone, ethanol) have been shown to inhibit the oxidation of different substrates by microsomal monoxygenases (Wolff, 1977). Caution is therefore advised in the selection of organic solvents and to reduce the risk of generating artefactual results, it is essential to use the minimum amount of solvent compatible with the adequate testing of the chemical under investigation.

Other solvents have been tested for their compatibility with bacterial mutation assays: Maron *et al.*, (1981) have presented a comprehensive study of the effect of different solvents on the performance of the *Salmonella* test. This paper should be consulted when solvents other than DMSO or water are to be used. Where DMSO or water are not used, it is recommended that the solvent chosen should be tested in trial runs at those concentrations to be used in the assay both in the presence and absence of S9-mix. The highest acceptable concentration of solvent is that which has no effect on the response of bacteria to a range of reference mutagens, and no effect on toxicity as judged by the microscopic examination of background lawn.

It is possible that a situation might arise where the test substance is totally insoluble in any solvent compatible with the test system. In these cases, the substance should be triturated, sonicated or emulsified to generate a fine suspension. A recent paper has shown that emulsification of insoluble hydrocarbons in pluronic polyol F127 might offer a useful method for adequately testing such compounds (Marino, 1987).

2.3 THE PROCEDURE

2.3.1 Outline of basic technique

The bacterial mutation assay which forms the basis of most if not all screening programmes is the *Salmonella* assay (also known as the 'Ames Test') (McCann *et al.*, 1975; McCann & Ames, 1976; Maron

& Ames, 1983). It is vital that workers contemplating the use of this assay should read and take note of the advice given in these references.

2.3.1.1 Pour-plate assays

These assays form the basic bacterial mutagen screening assay, as employed in most genotoxicity laboratories world-wide. The assay can either be performed using a standard plate-incorporation technique, or be extended by use of the pre-incubation modification of the assay.

(a) Standard plate-incorporation assays

The *Salmonella* assay employs the basic plate-incorporation technique: about 10^8 histidine-requiring bacteria (see Section 2.3.2.3), up to 0.1 ml of the test compound solution (or appropriate solvent or positive-control solutions), and 0.5 ml of S9-mix are added to 2 ml of molten 0.6% soft agar containing a trace of histidine and biotin. This mixture is then poured evenly across the surface of a base agar plate (containing glucose and a simple salts medium) and is allowed to set. The level of glucose in the base agar is normally 2%, but if TA97a is used routinely it has been recommended that this should be reduced to 1% (Piper & Kuzdas, 1987). High glucose concentrations inhibit the growth of strains which carry the his01242 mutation, such as TA97a, resulting in small revertant colonies. The other tester strains are not affected by this change in glucose concentration. The plates are then incubated (inverted, in the dark) at 37 °C for a period of 2–3 days (see Section 2.3.2.6(a)).

The trace of histidine in the top agar layer (usually 0.045 mM) allows the logarithmic growth (after a lag period, see Section 2.3.2.3) of the histidine-requiring bacteria in the presence of the test compound and/or any of the metabolites of the compound that are generated by the S9-mix. This period of several cell divisions is essential to allow the fixation of any pre-mutagenic lesions that have occurred in the bacterial DNA, before exhaustion of the histidine supply halts the growth of the auxotrophic cells. Only those cells which have been reverted to histidine-independence will continue to divide to form discrete, visible colonies randomly distributed across the test plate. The growth of the non-reverted cells forms a visible background lawn on the plate. Thinning or loss of this lawn is one non-quantitative indicator of compound-induced toxicity.

The basis of the assay is therefore to determine whether addition of graded doses of the test (or positive-control) compound to a series of such plates induces a dose-related increase in the number of observed mutant colonies compared with that obtained on the plates treated only with the appropriate volume of the solvent. As some test-chemicals are capable of direct interaction with DNA, the assay should be performed in both the absence and the presence of S9. Simple omission of the S9-mix component in the top agar is not recommended, as the differing volumes of the agar overlay will alter the perceived dose of compound (at least initially, depending on solubility and/or diffusion into the basal agar). The S9-mix component should be replaced with phosphate buffer. As no specific requirements are laid down in the various Regulatory Guidelines, it is largely a matter of practical convenience whether the treatments in the presence and absence of S9-mix are performed in the same or separate experiment(s). If quantitative comparisons are to be made between experiments carried out in the presence and absence of S9-mix, it is preferable to perform them on the same day.

(b) Pre-incubation tests

The use of the standard plate-incorporation protocol **with S9-mix** is known to be sub-optimal in the detection of a number of bacterial mutagens. These include aliphatic *N*-nitroso compounds (Bartsch *et al.*, 1976; Yahagi *et al.*, 1977), azo-dyes such as butter yellow (Bridges *et al.*, 1981, and individual reports in the same volume; Prival *et al.*, 1984), and alkaloids (Yamanaka *et al.*, 1979).

Such limitations of the standard protocol led to the development of a pre-incubation method where the bacteria, test-compound solution and S9-mix components are incubated (at temperatures between 30 °C and 37 °C) for various periods of time (e.g. 20 minutes: Yahagi *et al.*, 1975; 60 minutes: Lefevre & Ashby, 1981), before adding the soft agar and pouring as for the standard assay (Maron & Ames, 1983). Although it has been demonstrated that these apparent 'plate negative, pre-incubation positive' results are not always straightforward, e.g. with butter yellow (Robertson *et al.*, 1982a,b; Callander, 1986), the need for a pre-incubation step for the expression of some mutagens is recognised (e.g. Ashby *et al.*, 1985).

The National Toxicology Programme (NTP) has specifically adopted the pre-incubation protocol in preference to the standard plate assay (Haworth *et al.*, 1983; Zeiger *et al.*, 1988). Longer pre-incubation times (usually 60 minutes) generally increase the sensitivity of the assay to mutagens and offer advantages in convenience (Lefevre & Ashby, 1981; Gatehouse *et al.*, 1985). For allyl compounds pre-incubation times greater than 60 minutes together with aeration have been found necessary (Neudecker & Henschler, 1985a,b).

These results indicate that the use of the pre-incubation method (ideally using 60 minutes incubation at 37 °C) will undoubtably allow the detection of some indirect-acting mutagens that would be missed in the standard plate-incorporation assay. The question of how to accommodate the pre-incubation assay into the testing scheme in the most efficient way is considered in Section 2.3.2.5(c).

(c) Testing volatiles

For the special case of volatile compounds and gases, the plate-incorporation method is modified as follows (as quoted by Ames *et al.*, 1975): bacteria and S9 are mixed with top agar, and plated in the usual manner. The plates are then exposed, at 37 °C, to known atmospheric concentrations of the test compound in a sealed container equipped with a stirring system: time of exposure must be determined experimentally. After suitable exposure the atmosphere containing the test substance is replaced by air, and incubation is continued for the normal time (see Section 2.3.2.6(a)).

A more recent paper by Hughes *et al.*, (1987) describes a sensitive test method for the detection of mutagenic activity of volatile organic chemicals, using the *Salmonella*/microsome assay. This method involves the use of Tedlar bags to allow for the effective exposure of the bacterial test strain, and is particularly suitable for testing agents with boiling points <63 °C (e.g. ethylene oxide and methylene chloride). Hirota *et al.* (1987) have investigated the use of a 'bubbling procedure' for testing volatile compounds. Bridges (1978) has also studied the problems of testing volatile compounds and investigators are urged to consult these publications before attempting to test compounds of this nature.

2.3.1.2 Treat-and-plate tests

In all the above methods the test substance is applied to bacteria which are growing and dividing: this makes the test very sensitive but also makes it difficult to determine quantitatively the lethal, as opposed to the mutagenic, effects of the test substance. In certain cases (e.g. when feeding effects are suspected) it may be desirable to measure both effects simultaneously. This can be done by using the treat-and-plate method: bacteria are washed free of growth medium, resuspended in a suitable non-nutrient medium, and treated with the test substance, either in graded doses, or at a constant dose but varying the length of treatment. Separate samples of the treated bacteria are then plated on selective medium and on complete medium in order to determine the number of revertants and survivors respectively. This method is more useful for studying the mechanisms of mutagenesis with model mutagens than for routine screening, since it takes longer to perform than plate-incorporation assays. Green & Muriel (1976) give a comprehensive account of the treat-and-plate method, and pay special attention to the correct method of analysing data from such tests. The crucial point is that results may be incorrectly interpreted as positive when there is a treatment-related decrease in bacterial viability against a background of spontaneous plate mutants whose numbers are independent of the size (within wide limits) of the initial inoculum applied to the plate. It is essential, therefore, in calculating the frequency of mutation per survivor, to subtract the number of spontaneous revertants observed on untreated plates from the values obtained from treated plates before correcting for survival. Failure to observe this rule will lead to the sort of confusion predicted by Green & Muriel (1976), and drawn attention to by Venitt (1978) and Paes (1984).

The treat-and-plate method is specifically recommended by the Japanese Ministry of Health and Welfare Guidelines (JMHW, 1984), when highly toxic compounds, e.g. antibiotics, are being assayed for bacterial mutagenicity. This technique has also been used with the *E. coli* 343/113 system developed by Mohn and his colleagues (Mohn *et al.*, 1984), and Mitchell *et al.* (1980) have discussed the use of this method together with its inherent problems.

2.3.1.3 Fluctuation tests

The fluctuation test was originally designed by Luria & Delbruck (1943) to distinguish between mutation or adaption as the true explanation for bacterial variation. It has been used as a mutagenicity assay in its classic form by Voogd *et al.* (1974) and has been further developed by Green and his co-workers (Green *et al.*, 1976, 1977a,b; Green & Muriel, 1976), who simplified the method by adopting a modification suggested by Ryan (1955). Further modifications, using microtitration plates, have been introduced by Gatehouse (1978) and Gatehouse & Delow (1979). Detailed description of the methods employed are given in Hubbard *et al.* (1984).

The fluctuation test can detect mutagens which are effective only at near-lethal doses, and can distinguish very small increases in induced mutation compared with spontaneous rates. As well as its use in tests which determine reversion from auxotrophy to prototrophy in bacteria, it can also be used in a variety of organisms with any 'non-leaky' genetic end-point. For example, fluctuation tests can be used with yeast (Parry,

1977) and with cultured mammalian cells (Cole *et al.*, 1976; O'Neill *et al.*, 1981).

For reversion assays using, for example, *S. typhimurium* or *E. coli*, the test is performed as follows: A series of replicate incubation mixtures is prepared, containing, in a liquid minimal medium, at least 10^7 bacteria per ml, a trace amount of the essential amino acid, and where required, S9-mix. One of the incubation mixtures is reserved as a solvent (negative) control, and the rest receive graded doses of the test substance. Each incubation mixture is then immediately dispensed into, for example, 50 or 96 replicate vessels (small test-tubes or wells in plastic trays). Spontaneous or induced revertants arise during overnight incubation of these subdivided incubation mixtures in the auxotrophic phase of growth: the presence of one or more such revertants in each tube or well can be detected by further incubation (for 2–3 days), during which time revertants will continue to grow, even when the supply of the essential amino acid has been exhausted. The number of tubes or wells which contain revertants is conveniently scored by the addition of a suitable pH indicator dye, which changes colour in those tubes or wells where the pH has fallen in response to the florid growth of one or more revertants. A dose-related, statistically significant increase in the number of positive wells compared with appropriate negative controls is taken to indicate that the test substance is mutagenic. The mean number of revertants per well can be calculated from the zero term of the Poisson distribution (see, for example, Venitt, 1982). This parameter, rather than the number of positive wells, is often linearly related to dose. When testing coloured compounds or substances that interfere with the indicator dye it may be necessary to confirm growth by streaking out onto minimal agar (with biotin in the case of *Salmonella*).

The fluctuation test appears to offer metabolic advantages similar to the pre-incubation method. This may result from the fact that the concentration of cytosolic material remains constant, whereas in the pour-plate method the soluble enzymes diffuse into the bottom agar. This can lead to contrasting findings between the two methods for activation-dependent mutagens (Forster *et al.*, 1980, 1981; Gatehouse & Wedd, 1984).

The fluctuation test is susceptible to interference through feeding effects (Venitt & Bosworth, 1983) and more efficient use of nutrient materials (extender effects: Forster *et al.*, 1982) and this may be a particular problem for environmental samples or biologically derived materials as discussed earlier (Section 2.2.1). If necessary, the bacterial yield for each well or tube can be determined experimentally and the results can be corrected accordingly.

2.3.2 Critical factors in the procedure

2.3.2.1 Selection of appropriate organisms

In the previous edition of these guidelines (Venitt *et al.*, 1983) it was recommended that *Salmonella* strains TA1535, TA1537, TA1538, TA98 and TA100 be used for general screening and further recommended that *E. coli* WP2 *uvrA* (pKM101) should also be used. Since that time proposals have been made to revise the basic set of strains for screening (Maron & Ames, 1983), and two further strains TA97 and TA102 have been made available for use. Strain TA97 is a frameshift-detecting strain that has a + 1 frameshift mutation in a run of cytosines in the histidine D gene, and which also contains the pKM101 plasmid. Some of the characteristics of TA102 have been discussed previously (see Section 2.1.1). Basically, it has an AT base pair at the critical mutation site within the hisG gene, which is located on a multicopy plasmid pAQ1.

Another important feature of TA102 is that, unlike all the other recommended *Salmonella* strains, it has an intact excision repair system. This facilitates the detection of cross-linking agents, such as mitomycin C. Although it has been claimed that TA102 possesses certain unique features, some of these can be attained by the use of appropriate *E. coli* WP2 strains. For example, *E. coli* WP2 (pKM101) has an AT base pair at the critical mutation site, is excision proficient (and thus will detect cross-linking agents) and carries the pKM101 plasmid.

The use of a revised screening set of TA97, TA98, TA100 and TA102 has been advocated by Ames and his colleagues in terms of the wide coverage of genetic events afforded by this combination (Levin & Ames, 1986; Hartman *et al.*, 1986). The mutation sites of TA102 and TA100 together correspond to all possible base-substitution events (both transitions and all four transversions), while addition and deletion frameshifts can be detected in two different sequences in TA97 and TA98 (see Table 2.1).

The set of strains deployed must be justified not only in terms of genetic coverage but also in terms of efficiency in **screening** for mutagens and carcinogens. Since there is considerable overlap in the sensitivity of the tester strains (TA1535, TA1537, TA1538, TA98, TA100, TA97, TA102, WP2 *uvrA* (pKM101)), we may ask which strains may be excluded in the interests of economy. The arguments revolve around the issues of 'sensitivity' and the detection of mutagens rather than 'specificity' and the false-positive rate; the NTP data indicate that no particular strain had a predilection for yielding positive results with non-carcinogens (Zeiger, 1987).

It is generally accepted that the combination of TA98 and TA100 is very sensitive, detecting a large proportion of known bacterial mutagens. Arguments for retaining the equivalent non-plasmid strains are usually based on the incidence of mutagens which are active only in these strains (Bridges *et al.*, 1981). 'Unique' (observed in one strain only) positive results in TA1535 are easily found in several published data-sets (Bridges *et al.*, 1981; Haworth *et al.*, 1983; Moriya *et al.*, 1983; De Flora *et al.*, 1984b; Zeiger & Haworth, 1985; Zeiger, 1987; Ashby & Tennant, 1988). Examples of unique positives with TA1538 are more difficult to find (the extensive NTP data-base does not include TA1538). In an analysis of results from 1221 tests, TA1538 was found to give a lower incidence of unique positives than the other strains (Herbold, 1982). No loss in sensitivity was noted when TA1538 was dropped from a screening set (Dyrby & Ingvardsen, 1983), and the redundancy of this strain has been recognised by groups of experts of the OECD, Gene-Tox program and US EPA (OECD, 1983; Kier *et al.*, 1986).

It has been suggested that TA1537 should be substituted by TA97 (which also overlaps in sensitivity with the strains containing the D3052 mutation, TA98 and TA1538). There are few data-bases on which to make a comparison, and the best is probably the original publication from Ames's group (Levin *et al.*, 1982b), in which TA97 is shown to detect several mutagens not detected by TA1537 (substituted triazines, phenothiazines, PR toxin) and to show greater sensitivity to several others. Other examples of unique positive and/or the greater sensitivity in TA97 can be found in the literature (De Flora *et al.*, 1984a; MacGregor & Wilson, 1985; Sakai *et al.*, 1985; Zeiger, 1987; Kappas, 1988), but some counter examples of compounds to which TA1537 is more sensitive can also be found (Brown & Brown, 1976; Thomas & McPhee, 1984; De Flora *et al.*, 1984a; Bonneau & Cordier, 1985; Zeiger, 1987; M.R. O'Donovan, personal communication). For one series of related mutagenic compounds Shahin *et al.* (1985) found that the fold increase in revertants over control levels was greater with TA1537, but that more revertants per nanomole were obtained with TA97. Information is not available from an extensive data-base to indicate the frequency with which TA1537 detects mutagens to which TA97 is not sensitive, and vice versa.

The case for the inclusion of TA102 is based on its demonstrated sensitivity to classes of compounds not detected or poorly detected by other tester strains (Levin *et al.*, 1982a; De Flora *et al.*, 1984b). These are (i) quinones and redox-cyclers, (ii) hydroperoxides, (iii) aldehydes and (iv) cross-linking agents, e.g. mitomycin C. Thus, in addition to the wide genetic coverage offered by this strain, the increased coverage of chemical classes would seem to justify its use. However, little further work with this strain is reported in the literature to permit assessment of these claims. In one study of 20 peroxides and aldehydes tested in TA100 and TA102, none was uniquely positive in TA102, and responses in TA102 were always accompanied by weak responses in TA100 (E. Zeiger, personal communication).

The plasmid-containing *E. coli* strain WP2 series appears to be a useful group of supplementary strains (McMahon *et al.*, 1979; Venitt & Croften-Sleigh, 1981; Dyrby & Ingvardsen, 1983) which can be used with the convenient pour-plate method, in the same way as *Salmonella* strains. The WP2 strains provide data on the reversion of an auxotrophic end-point, in the context of a genetic background other than *Salmonella*. *E. coli* WP2 *uvrA* (pKM101) has been demonstrated to detect carcinogens with the 'accuracy' of the TA98 and TA100 combination (Venitt & Crofton-Sleigh, 1981). The use of a strain from the *E. coli* WP2 series is specifically requested by the Japanese MHW Guidelines (1984). The tryptophan mutation in the WP2 series is a chromosomal 'ochre' mutation, and the strains should therefore parallel the *Salmonella* tester strain TA104, although few data are available from the screening of chemicals to support this contention. Wilcox *et al.* (1990) showed that a combination of a repair-proficient and a repair-deficient *E. coli* strain allowed the detection of a similar range of oxidative mutagens and cross-linking agents as identified by strain TA102.

The essential advantages of speed and simplicity are offered by all the tester strains, old and new. There have been some initial difficulties with the maintenance of TA97 and TA102. The presence of slow growing suppressor mutants and variations in the copy number of pAQ1 appear to reduce the reproducibility of TA102 (Goggleman & Vollmar, 1985; Grafe & Goggleman, 1985; Albertini & Gocke, 1988). It was necessary to reconstruct TA97 which was re-released as TA97a with a recommendation to use lower glucose levels in the agar plates (Piper & Kuzdas, 1987). This re-release was not accompanied by a formal publication of supporting data from Professor Ames's group. The strain also contains a temperature sensitive operator constitutive mutation (his01242) and if incubation temperatures exceed 40 °C this mutation may cause variability in background mutation frequencies, and morphological changes.

Neither TA97 (TA97a) nor TA102 have been widely adopted and have not enjoyed the same popularity as the previous *Salmonella* tester strains. This point is illustrated by the relative paucity of publications which have appeared making use of these strains. This may in part result from the difficulties which have been experienced by many laboratories in maintaining them. Difficulties with strain maintenance were also experienced by many investigators in the early days of TA98 and TA100 (e.g. Bridges *et al.*, 1981), and led to several studies of interlaboratory reproducibility and genetic drift (Margolin *et al.*, 1983). Similar studies have not yet been undertaken with TA97a and TA102, but would be valuable in illustrating the robustness and reliability of the strains in different laboratories and in providing reference data for the correct use and performance of these strains.

In summary, the limited data available suggest that TA97 and TA102 should offer significant advantages for the efficient screening of chemicals as well as theoretical advantages in providing a more comprehensive coverage of genetic events. It is recognised, however, that there have been difficulties in the maintenance of the strains, and there are limited published data on their performance. On the basis of the arguments presented above, it would be desirable to retain TA1535 in the basic screening set, although TA1538 may be excluded. *E. coli* offers several advantages for routine screening, in particular a genetic background which is different from *Salmonella typhimurium*, demonstrated sensitivity, and a similar mutation site to strains TA102/TA104. It is proposed that routine screening should be performed using *S. typhimurium* strains TA1535, TA1537, TA98, TA100 and TA102. For the reasons outlined above the *E. coli* strains WP2 *uvr*A (pKM101) and WP2 (pKM101), in combination, are acceptable as an alternative to *S. typhimurium* strain TA102. It is also proposed that strain TA97a can be used an alternative to strain TA1537. However, the working group strongly advocates that additional comparative studies should be carried out on the latter two strains to substantiate this proposal further.

Testing strategies based on these combinations of strains should be considered of equal validity. It is recognised that even the most comprehensive screening battery will not detect every mutagen, but the battery given above is believed to be one which will make the most effective use of time and resources. This proposal has a minimal impact on the workload for the testing of chemicals.

It may be useful to use alternatives to this combination of strains for specific reasons: for example, when testing hydrazines, strains TA1530 and hisG46 have been shown to be more sensitive than corresponding cell wall deficient strains (Tosk *et al.*, 1979). In specific circumstances when investigating the possible role of bacterial metabolism in the observed mutagenic effects, the use of nitroreductase-deficient (Rosenkranz *et al.*, 1982) and glutathione-deficient strains (Kerklaan *et al.*, 1985) should be considered. When testing foodstuffs it may be desirable to employ a strain which uses drug-resistance as its genetic end-point. This would avoid the possible confounding effects of contaminating amino acids in the test substance (discussed in Section 2.2.1).

2.3.2.2 Genetic stability of test system

There is no substantive evidence that the bacterial strains routinely used for screening are genetically unstable.

The conclusion drawn from a large collaborative study was that genetic drift was a minor component in interlaboratory variability (Margolin *et al.*, 1983). These authors emphasised, however, that the role of genetic drift is minimised by regular checking of the cultures and rejection of those not showing the appropriate properties. Reports of the high variability of the phenotypic characters of the *S. typhimurium* strains (Speck *et al.*, 1975; Anders *et al.*, 1982) were subsequently criticised (McPhee, 1984; Claxton *et al.*, 1984). It should be noted that there may be selection for some attributes (glutathione export, loss of H_2S production) during the long-term maintenance of stock cultures (Hartman, 1987). A protocol for the combined biochemical and serological identification of the Ames *S. typhimurium* tester strains has been published recently by Busch *et al.* (1986).

It is of paramount importance that tester strains should be regularly checked for possession of all their characteristic genetic features, and stored under appropriate conditions. It has been recommended that strain TA102 should be maintained in media containing ampicillin and tetracycline to preserve the plasmid copy number (Albertini & Gocke, 1988). The properties to be checked should include the following:

1. amino acid requirement (e.g. histidine for *S. typhimurium*, tryptophan for *E. coli* WP2 strains);
2. background mutation rates and induced mutation rates with reference mutagens;
3. presence of R-factor plasmids where appropriate (e.g. ampicillin-resistance in TA98, TA100 and TA97, WP2 *uvrA* (pKM101), WP2 (pKM101), and ampicillin- and tetracycline-resistance with TA102);
4. presence of characteristic mutations (e.g. *rfa* and *uvrB* mutations in *S. typhimurium* strains, *uvrA* mutation in *E. coli* strains).

Ames *et al.* (1975) provide detailed instructions for the maintenance and checking of strains used in the *Salmonella* test. The frequency of such checks is a matter of choice: some checks (3, 4) may be carried out at weekly intervals, others (1, 2) can be incorporated into experimental design. For example, much information may be gleaned from the response of tester strains to reference mutagens ('positive controls') run concurrently in each experiment, and from the spontaneous mutation rates seen on untreated or solvent-treated plates ('negative controls') (see Section 3.2.4.1). Relevant data on the characterisation of the tester strains used in a given assay must be available for scrutiny.

Permanent master-cultures of tester strains should be stored in liquid nitrogen or below −70 °C. These master-cultures should be prepared and checked by the appropriate method before storage. Working cultures for use in screening experiments should be prepared by inoculation from a master-culture or from a plate made from a master-culture – never by passage from a previously used working culture. Passage of tester strains from one working culture to another will inevitably increase the number of pre-existing mutants, leading to unacceptably high spontaneous mutation rates.

2.3.2.3 Culture conditions

There has been much discussion about the best way to prepare cultures for use in plate incorporation tests. Ames *et al.* (1975) recommended the use of overnight cultures in nutrient broth: a group of investigators with considerable experience of bacterial testing endorsed this method (De Serres & Shelby, 1979). The incubation period was later revised to 10 hours, since it was shown that viability decreases in most nutrient broth cultures grown longer than 12 hours (Maron & Ames, 1983). Other workers claim that the use of washed logarithmic-phase cultures increases the sensitivity of the standard *Salmonella* test (e.g. Booth *et al.*, 1980). There is some evidence that the use of log-phase bacteria can increase their response to the mutagenic effects of alkylating agents (Hince & Neale, 1977), and can influence the response to compounds which are detoxified by intracellular glutathione (Goggleman, 1980). A possible reason for increased sensitivity with log-phase cells is that the lag-period before the bacteria begin to replicate on the plate (see below) is shortened if the plates are inoculated with actively growing cells (Wedd *et al.*, 1988).

It has been argued previously that because the agar overlay is supplemented with trace histidine and biotin, the bacteria will replicate in the agar when the plates are incubated and hence the starting condition of the bacteria is not critical. The fact that the bacteria are actively growing in the presence of the test agent is indeed an important feature of the plate-incorporation assay since it provides an opportunity for DNA damage to be induced during DNA replication, increasing the probability that mutations will occur (Venitt *et al.*, 1983). However, it has been shown that, after the bacteria are inoculated into the soft agar overlay, there is a lag-period of around 4–6 hours prior to active cell replication (Barber *et al.*, 1983). The existence of this lag-phase could have a marked effect on the magnitude of response obtained with (a) highly labile, direct-acting mutagens which have a short half-life in the agar medium, (b) replication-dependant mutagens, (c) compounds which are converted to labile mutagenic metabolites by S9-mix (S9 enzymes appear to lose activity in the agar plates after only a few hours). Barber *et al.* (1983) illustrated this point using a delayed-plating method, whereby the bacterial cells are plated into the top agar and incubated at 37 °C for 6 hours before applying a second agar layer containing the test agent. Using this technique, the bacteria are already replicating when they are exposed to the test agent, and Barber *et al.*, were able to demonstrate a greatly enhanced response with MNNG (TA1535−S9) 9-aminoacridine (TA1537 −S9) and 2-aminoanthracene (TA1535 and TA100 +S9).

While useful for illustrating the importance of the lag-phase, the delayed-plating method is too labour-intensive for routine screening. It is, however, recommended as a useful protocol variation if ambiguous results are obtained in the standard assay or if the test agent is suspected of being unstable under the conditions of the assay.

There is one aspect of the bacterial culture which is, however, critical: it is very important that the culture contains sufficient organisms to ensure that each plate receives at least 10^8 viable bacteria. In order to detect induced mutation frequencies of the order of 1 in 10^6, it is necessary to ensure that the bacterial population is well in excess of this size by the end of the period of auxotrophic growth. Since the mutagen or its metabolites are likely to be most active during the first few hours of incubation, a large number of bacteria (ideally exponentially growing cells) should be present at the outset. Thus, it is essential that the cultures used in plate tests contain a high titre of viable bacteria and this should be demonstrated as part of each assay by performing a viable count on the stock culture. To detect only weakly mutagenic chemicals it may prove beneficial to expose even larger numbers of bacteria to the test agent. This approach has been employed by Kado *et al.* (1983), who exposed 10^9 cells to urine concentrates from mutagen-treated animals and reported a 20-fold increase in sensitivity when compared to the standard plate-incorporation assay and a 13-fold increase over the pre-incubation test. For further discussion on the effect of inoculum size and growth rate of the plated bacteria on the yield of spontaneous and induced revertants in the plate incorporation test, the reader is referred to papers by Salmeen Durisin (1981) and Goggleman *et al.* (1983).

2.3.2.4 Controls and internal monitoring

Each assay should include the following controls:

1. Negative controls, to which the solvent vehicle, but no test substance, has been added. (When solvents other than water are used, a concurrent 'untreated' control made up to the same volume with buffer, should be included in every experiment.)
2. Positive controls: these should be reference mutagens to check that the assay is performing correctly. Appropriate mutagens are listed by De Serres & Shelby (1979), Maron & Ames (1983) and Venitt *et al.* (1984). Pagano & Zeiger (1985) have recently shown that it is possible to store stock solutions of some of the most routinely used positive controls (sodium azide, B(a)P, 2-aminoanthracene, 4-nitroquinoline oxide) at −20 °C to −80 °C without loss of activity, thus reducing potential exposure of laboratory personnel.

(a) Negative controls

The observed number of mutant colonies on untreated or solvent-treated plates for each tester strain is a useful internal monitor for each assay. This control value (frequently mis-termed the 'spontaneous mutation rate') in a plate-incorporation test depends, *inter alia*, on (i) the intrinsic mutability of the given strain (e.g. plasmid-bearing strains are more mutable than non-plasmid-bearing strains), (ii) the total number of cell divisions of the bacteria on the plate (which depends on the supply of nutrients, especially histidine and biotin in *S. typhimurium*), (iii) the number of mutants pre-existing in the culture which have been added to the plate. Factors (i) and (ii) govern the number of 'plate mutants' – those mutants which arise during the phase of auxotrophic cell division on the plate. The presence of S9, which contains trace amounts of nutrients, also has a small effect upon the number of plate mutants, usually increasing it (Table 2.2). Factor (iii) depends on the total number of cell divisions of the bacteria during their period of growth in broth, the size of the initial inoculum and its own burden of pre-existing mutants, the time during the growth of the broth culture at which mutations occur, and the number of broth-grown bacteria added to the plate. In other words, cultures will inevitably contain varying numbers of revertants which have arisen during incubation in nutrient broth. The interplay of these different factors accounts to a large extent for the wide range

Table 2.2. *Acceptable ranges of background revertant counts for the routinely used strains*

Test strain	Range of background colonies per plate	Reference
TA1535	3–37	Kier *et al.*, 1986
TA1537	4–31	Kier *et al.*, 1986
TA98	15–60	Kier *et al.*, 1986
TA100	75–200	Kier *et al.*, 1986
TA97 (TA97a)	90–180	Maron & Ames, 1983
TA102	240–360	Ames (personal comm.)
WP2 uvrA (pKM101)	45–151	Venitt *et al.*, 1984
WP2 (pKM101)	35–160	Wilcox *et al.*, 1990

of background mutation rates reported for the frequently used *Salmonella* strains in a number of collaborative studies (De Serres & Shelby, 1979; De Serres & Ashby, 1981; Chu *et al.*, 1981; Venitt, 1982). A detailed discussion of these factors can be found in Salmeen & Durisin (1981).

Despite these problems the control mutation values (revertant colony plate counts) do provide an indication that the tester strains are being adequately maintained and the assay is being properly conducted. Each laboratory should determine the normal range of control revertant colonies per plate for each strain. The range might vary between laboratories but those given in Table 2.2 should be taken as a guide.

Sudden deviations in background reversion counts, even if they fall within the acceptable range given in Table 2.2, should be investigated, and those assays where such deviations have occurred should be repeated in experiments where spontaneous reversion rates show a consistent value.

In addition to these negative controls, frequent checks should be made on the sterility of S9 preparations, media and plates. Whether plates are prepared 'in-house' or are bought in as proprietary ready-poured plates, they should be dry enough to prevent premature coalescence of colonies, but not old, cracked and wrinkled. Special attention should be paid to the volume of agar per 90-mm plate, since experience has shown that the use of thin plates, containing less than the 25 ml agar recommended by Ames *et al.* (1975), can reduce the yield of revertants.

(b) Positive controls

The correct use of positive control chemicals has been discussed in general in Chapter 1.

(c) Assessment of toxicity

In addition to these internal controls, the background lawn of samples of both treated and control plates should be inspected microscopically in order to check for toxic effects (thinning of the lawn), or for excess growth which may indicate the presence of histidine or histidine precursors in the test material. There is no convenient, well tried and rapid way of quantifying the extent of growth of the background lawn, but visual examination gives a useful impression of its quality.

Pronounced toxicity may reduce the population of viable bacteria to such an extent that only a few hundred bacteria will survive. Auxotrophic cells will continue to grow because competition for the available supplementation reduces with increasing toxicity. These cells may each form a small visible colony, rather than coalescing into a confluent lawn, and these 'micro-colonies' may be mis-interpreted as mutant colonies by inexperienced workers. In cases where it is suspected that colonies seen on treated plates are not true revertants, for this or any other reason (e.g. formation of phenocopies, surfactant effects) replica plating to minimal plates (supplemented with biotin in the case of *S. typhimurium*) should be undertaken to establish the proportion of true revertants. Such checks do not normally form part of a minimum protocol, but should certainly be undertaken if there is any suspicion about the results of a test.

2.3.2.5 *Metabolic activation*

(a) Sources of metabolising systems

The preparation routinely used is a $9000 \times g$ supernatant (S9 fraction) of liver obtained from rats pretreated with Aroclor 1254 to induce drug-metabolising enzymes. Such a preparation has been shown to allow detection of a wide range of promutagens (Ames *et al.*, 1975; McCann *et al.*, 1975; McCann & Ames, 1976; IARC, 1980b; De Serres & Ashby, 1981).

Concern about the toxicity, carcinogenicity and persistant nature of Aroclor 1254 in the environment has encouraged searches for safer enzyme inducers. The use of a combination of phenobarbitone and β-naphthoflavone (5,6-benzoflavone) results in 'mixed' induction of cytochromes P450-dependent drug metabolising enzyme activities similar to that seen after Aroclor exposure (Ong *et al.*, 1980; Suter & Jaeger, 1981). S9 prepared from phenobarbitone/β-naphthoflavone treated rats has been shown to activate mutagens from a variety of chemical classes (Matsushima *et al.*, 1978; Gatehouse & Delow, 1979; Ong *et al.*, 1980) and so would appear to be a satisfactory substitute for Aroclor-induced

S9, although the number of promutagens studied is far greater with the latter preparation. It should be noted that PB/BNF combination is the preferred inducer under the guidelines issued by Japanese authorities (Ishidate, 1988).

In comparative studies using coded compounds, Syrian hamster or mouse S9 (induced or uninduced) offered no overall advantage over Aroclor-induced rat S9 for detection of mutagens (Haworth *et al.*, 1983; Dunkel *et al.*, 1984). However, on occasions it may be more appropriate to use alternatives to the usual Aroclor/rat liver S9 preparation, particularly when screening chemicals of classes known to be detected more efficiently using other metabolising systems. For example hamster S9 activates many aromatic amines, heterocyclic amines, N-nitrosamines and azo-dyes (Prival & Mitchell, 1982) more effectively than rat S9 (Lijinsky & Andrews, 1983; Alldrick & Rowland, 1985; Le *et al.*, 1985; Hyde *et al.*, 1987; Zeiger *et al.*, 1988). Hamster S9 is also essential for the demonstration of the mutagenic activity of phenacetin and other compounds (Zeiger *et al.*, 1988). Although the species differences are often quantitative, a number of N-nitrosamines and substituted anilines were detectable in the presence of hamster but not rat S9. It should be noted that the activating capacity of hamster S9, unlike that of rats, is not usually increased by pre-treatment of the animals with Aroclor (Alldrick & Rowland, 1985; Hyde *et al.*, 1987). The use of S9 fraction from extra-hepatic organs is generally of little use for studying organotropic effects of procarcinogens (or predicting them).

There is limited evidence to suggest that the activating capacity of S9 decreases as the age of the donor animal increases (Raineri *et al.*, 1986; Brennan-Craddock *et al.*, 1987) so it would seem prudent to use young (6–8 week old) male animals, as recommended by Ames *et al.* (1975). Comparative studies of S9 fraction and hepatocytes indicate differences in activating capacity between homogenates and intact cells (Bos *et al.*, 1983; Rumruen & Pool, 1984; Hass *et al.*, 1986). However, hepatocytes did not consistently improve the ability of the *Salmonella* assay to detect mutagens and so are not appropriate for general screening purposes. Nevertheless they clearly have an important place in further investigations on mutagens and can aid the interpretation of results and extrapolation of *in vitro* data to the intact animals (Hass *et al.*, 1986).

It is obvious from the above that manipulation of the *in vitro* metabolic activation system offers the easiest route to the manipulation of the outcome of bacterial mutation tests. Accordingly, any modification of the base-line protocol should be supported by sound scientific arguments together with details of the species, sex, strain and source of the animals used, including their diet, age, weight and inducing treatment.

Where human exposure occurs via the digestive tract, materials such as drugs and food additives are exposed to a number of processes which may modify their biological effects. Such modifying factors include exposure to digestive enzymes, nitrosation in the presence of nitrite and bacteria, and the complex metabolic activity of the gut microflora. Several systems have been developed which attempt to simulate these effects *in vitro* (e.g. McCoy *et al.*, 1977; Coulston & Dunne, 1979; Tamura *et al.*, 1980; Phillips *et al.*, 1980; De Flora & Picciotto, 1980; Karpinsky & Rosenkranz, 1980; Cerniglia *et al.*, 1986). These systems, although of great value in research programmes, have not received sufficient independent validation to allow their recommendation for general screening purposes. However, where the identity of the test compound is known, the use of additional activating systems may be appropriate. The exogenous metabolic activation system should be designed so as to provide the most likely chance of detecting potential activity.

The above discussion serves to underline the importance of knowing the identity of the test material so that the most appropriate activating conditions can be chosen on the basis of sound scientific judgment.

(b) *Preparation, characterisation and storage of hepatic S9*

Detailed discussion on the preparation of S9 is given in Chapter 1.

(c) *The S9-mix*

Most laboratories prepare S9 from a 25% liver homogenate (1 g liver plus 3 g buffer) and use 0.08 ml–0.1 ml of this S9 per ml of S9-mix (often termed 8–10% S9) for their general screening procedure. However, since different laboratories use different ratios of tissue to homogenising fluid during S9 preparations, and may also use different concentrations of S9 and co-factors in the S9-mix, it is necessary to specify in detail the method used for preparing and using S9 or to express the concentration of homogenate as mg liver protein per plate.

The activity of mixed function oxidases (MFO) in S9 is dependent on a continuous supply of NADPH generated from added NADP by endogenous glucose-6-phosphate dehydrogenase acting on added glucose-6-phosphate. For some compounds, e.g. aflatoxin B1 and cyclophosphamide, increased responses can be obtained by adding more NADP (Booth *et al.*, 1980). It has also been suggested recently that because the activity of endogenous isocitrate dehydrogenase is higher than that of glucose-6-phosphate in rat hepatic S9, isocitrate should be used instead of glucose-6-phosphate in the S9-mix to promote higher MFO activity (Lindblad & Jackim, 1982). Addition of flavin mononucleotide (FMN) to the S9-mix has been shown to facilitate the detection of some azo-dyes (Prival *et al.*, 1984, 1986). In this modified procedure, untreated hamsters are the source of the S9-fraction (rather than induced rats). The S9-fraction, together with FMN, exogenous glucose-6-phosphate dehydrogenase, NADH and an excess of glucose-6-phosphate (EPA, 1984), are pre-incubated with the test bacteria prior to addition to the top agar. This procedure allows effective azo-reduction to occur, enhances the mutagenic response observed for benzidine and is required in some cases to obtain a mutagenic response with certain benzidine-based dyes. However, further data are required before it can be recommended as a general approach to testing such compounds.

The modification of co-factor concentration and type in the S9-mix is worth considering when trying to resolve borderline results obtained using a conventional assay. For example addition of acetyl coenzyme A has been shown to potentiate the response with aromatic amines such as benzidine (Kennelly *et al.*, 1984).

Just as no single tissue preparation is adequate for the detection of all classes of chemical mutagens, no single concentration of S9 in the S9-mix will detect all these different classes with equal efficiency. The optimum concentration of the S9 in the S9-mix for any given compound or chemical class cannot be predicted by theoretical arguments, neither can past experience be used to judge the best concentration required for the activation of novel compounds. Ames *et al.* (1975) noted that 'too much as well as too little S9 can drastically lower the sensitivity'. Thus misleading results could be obtained if only one S9 concentration were used and several workers have recommended either two concentrations (4 and 10%) of S9 (Maron & Ames, 1983) or three concentrations (4, 10 and 30%) of S9 (Venitt *et al.*, 1984; Ashby, 1986). It is certainly the case that some mutagens (e.g. dimethylaminoazobenzene) can be more efficiently detected in the presence of high (>10%) concentrations of S9 (Callander, 1986). Petroleum-derived complex mixtures containing polycyclic aromatic hydrocarbons are activated to mutagens more efficiently when 2–8 times the normal amount of S9 is used (Carver *et al.*, 1985). The case for using a low concentration of S9-mix (e.g. 4%) is more difficult to justify. The Working Group was not aware of any mutagen which would invoke a response using 4% S9-mix but not with 8–10% S9-mix. The activity of certain aromatic amines, e.g.

2-aminoanthracene, decreases as the S9 concentration is increased (Callander, 1986), but it is readily detectable using 10% S9-mix.

A practical compromise to carrying out tests using several levels of S9 would to be use 8–10% S9-mix in both the plate-incorporation and pre-incubation tests. If 10% S9-mix is used in the plate-incorporation assay the concentration of S9 fraction in the top agar is 1.8% (50 μl in a total of 2.7 ml). However, because of the smaller treatment volume, the use of 10% S9-mix in the pre-incubation assay results in an S9 concentration of 7.1% (50 μl in 0.7 ml) during the critical period, i.e. the liquid pre-incubation phase.

It is recommended that a compound should first be tested in the plate-incorporation assay and then, if negative results are obtained, the pre-incubation assay. This strategy, which is summarised in Fig. 2.1, provides two types of exposure conditions (agar incorporation and liquid incubation), two levels of S9 fraction (1.8% and 7.1% final concentration) and some degree of repetition of experiments. (NB The lack of exact repetition of the pour-plate assay with S9-mix is not compliant with current OECD recommendations, 1983.)

2.3.2.6 *Incubation time and scoring of revertant colonies*

(a) Incubation conditions

Plates should be incubated in the dark in incubators which do not suffer from temperature gradients. Stacking of plates too tightly may also lead to uneven heat distribution. This is another reason for using incubation times longer than 48 hours (see Section (b) below). A reasonable policy would be to employ several well calibrated incubators to spread the load, rather than one large incubator crammed full of closely packed plates. It is extremely important to ensure that volatile test compounds and gases are prevented from leaking into positive and negative control plates or to other experiments in the same incubator. The sealing of plates in polythene bags or gas jars may be appropriate. The use of fan-assisted incubators to overcome these problems can lead to other problems, such as contamination of plates and the accumulation of moisture, and is not recommended.

(b) Time of scoring

Ames *et al.* (1975) recommend that in the standard *Salmonella* test, plates should be scored after 48 hours' incubation. However, experience has shown that the toxicity of some chemicals may retard the growth of revertants, and that the appearance of slowly growing suppressor mutants may also be delayed. Incubation for greater than 48 hours has

48 hours has been recommended when testing complex mixtures, since it has been shown that masking of mutagenicity by toxic chemicals present in such mixtures can occur if a 48 hour incubation time is employed (Zeiger & Pagano, 1984). Thus, it has been suggested that 72 hours is a more suitable period (De Serres & Shelby, 1979).

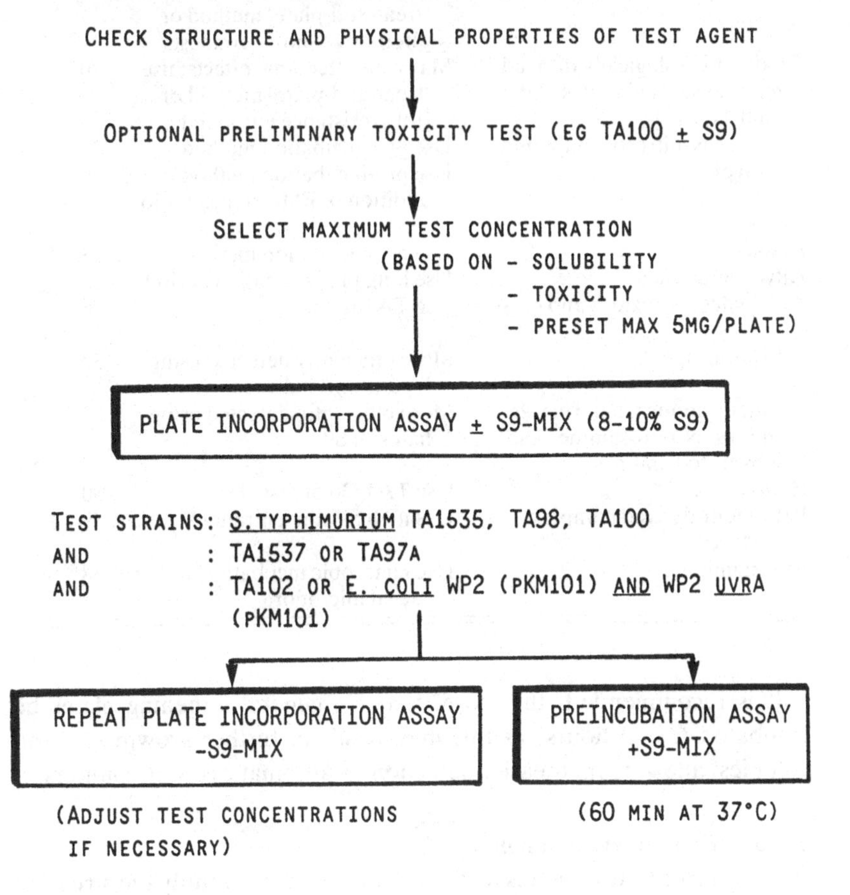

NOTE

1 SCHEME ASSUMES NO EVIDENCE OF ACTIVITY THROUGHOUT.
2 IF POSITIVE/AMBIGUOUS RESULTS OBTAINED AT ANY STAGE PROTOCOL SHOULD BE MODIFIED TO CONFIRM/EXTEND RESULTS.
3 IF A SIMILARITY TO A KNOWN CHEMICAL CARCINOGEN CLASS IS APPARENT THEN, IF POSSIBLE, AN APPROPRIATE POSITIVE CONTROL SHOULD BE INCLUDED, AND THE TESTING CONDITIONS GIVEN IN TABLE 3 SHOULD BE USED.

Fig. 2.1. Bacterial mutation assays – minimal testing scheme

Table 2.3. *Recommendations for specific classes of compounds*

Class of compound	Limitation or recommendation	Page no.
Antibiotics, surfactants, biocides and preservatives	May present special problems due to bactericidal activity. Use 'treat-and-plate' method or alternative short-term tests	19
Foods and biologically derived materials, also indoles and imidazoles	May cause 'feeding' effects; use 'treat-and-plate' method or drug resistance end-point	20
Aliphatic N-nitroso compounds	Use pre-incubation method	23
Azo-dyes	Use pre-incubation method, addition of FMN to incubation mixture	39
Alkaloids	Use pre-incubation method	23
Allyl compounds	Use long pre-incubation period	23
Aldehydes, peroxides and cross-linkers	Use TA102	28
Alkylating agents	More efficiently detected using logarithmic cultures	32
Aromatic amines, heterocyclic amines, N-nitrosamines and substituted anilines	More efficiently detected using hamster S9	37
Hydrazines	Use TA1530 or hisG46	30
Petroleum-derived complex mixtures	Use high S9 concentrations	39
Faecal mutagens	Use anaerobic incubation and caecal microflora	38

It is recommended, therefore, that for general screening plates be incubated for 72 hours, as this may result in further growth of small colonies, allowing their easier detection by automatic colony counters.

2.3.3 Experimental design

Prior to testing it is important to consider carefully the structure and physical properties of the chemical for clues to the appropriate selection of tester strains or conditions, on the basis of published results with analogous compounds. Examples of recommendations for specific classes of compounds are given in Table 2.3. These examples are drawn from the text of the present chapter.

The experimental design for bacterial mutation assays has been discussed at length in UKEMS Guidelines Part III, Chapter 2 (Mahon *et al.*, 1989). In this chapter definite recommendations have been made with regard to replication of treatment groups, details on dose ranges, dose intervals and the need to repeat experiments. These will be discussed briefly in the following paragraphs. However, for a detailed presentation of the statistical arguments underlying these recommendations, the reader is referred to the original document.

2.3.3.1 Number of replicate treatments

It is recommended that at least three plates are used per treatment, with at least five treatments + zero-treatment (vehicle) controls. Duplicate plates are sufficient for positive-control treatments only. These are the minimum requirements for a single assay. This contradicts the recommendations of Maron & Ames (1983), where duplicate plates per dose of test compound are considered adequate, but is consistent with the OECD Guidelines (1983). The use of twice as many negative control plates will lead to more powerful tests, and is recommended.

2.3.3.2 Number of treatment levels and intervals between treatments

Two- to three-fold concentration intervals are recommended, going downwards from the toxicity, solubility or upper limit. Experience has shown that doses which differ by factors of less than 10 should be used for the initial screening of compounds (Bridges *et al.*, 1981) to avoid the occurrence of false-negative results due to 'window' effects (e.g. Kathon biocide: Scribner *et al.*, 1983).

At least five treatment levels should be tested for each compound to allow for the establishment of a dose-response curve (OECD, 1983; JMHW, 1984).

2.3.3.3 Upper limits for testing

Many guidelines recommend as one of their minimum criteria for bacterial assays that compounds should be tested '. . . to the limit of toxicity or solubility . . .' (solubility is defined as the maximum concentration attainable within the most appropriate solvent *after* dilution in the test system). In practice, this may lead to some absurdities and the investigator may wish to use some discretion with those compounds which fail to reach either limit at extremely high doses. Nonetheless, it is important to test well into the milligrams range and a maximum test concentration of 5 mg per plate is recommended, since it has been shown that certain mutagens are active in plate-incorporation tests only at high concentrations (for example, N-nitrosomorpholine, 1-naphthylamine, cyclosphosphamide: see various Investigators' Reports in De Serres & Ashby, 1981).

When very high doses are tested account must be taken of the possibility of impurities in the test substance which may themselves contribute to toxic or mutagenic effects. It may, on occasions, be desirable to test concentrations giving rise to visible precipitation in the test system, as there are examples of compounds which induce increased revertant numbers only at test concentrations producing heavy precipitation on the agar plate (e.g. dimethylaminoazobenzene; D & C Red No. 9).

Toxicity has been defined in terms of reduced revertant numbers or thinning of background lawn; where testing is limited by toxicity, at the high dose-level there should be a moderate level of toxicity as indicated by at least one of these measures. Alternative methods are available for assessment of cell killing using 'filler' cells (Waleh *et al.*, 1982) and of background lawn growth by microscopic evaluation (McGregor *et al.*, 1984). It has been shown that these two measures of toxicity sometimes act independently (McGregor & Prentice, 1985). It is to be hoped that researchers will adopt these methods and attempt to obtain a clear understanding of the relationship between concurrent toxicity and mutation induction, since this remains a weak point of the Ames test.

2.3.3.4 Number of independent experiments and extent of testing

It is recommended that all experiments should be repeated at least once. Repeat testing is an effective way of reducing statistical errors (false positive or false negative).

The design of the second experiment depends on the results of the first experiment. If a strongly positive result is obtained, this should be confirmed in a second experiment designed to explore the linear part of the dose-response curve.

When negative results are obtained the scheme outlined previously in Fig. 2.1 is recommended. This provides for some duplication of experiments but also extends the range of conditions available for metabolic activation for tests carried out in the presence of S9-mix.

If any experiment produces a weak or equivocal result (e.g. statistically significant increases at isolated doses), the experiment should be repeated until a consistent picture emerges, for example, by using a narrower dose range so as to maximise the chance of obtaining a dose-response. Additional modifications to the protocol, as discussed earlier, should be considered. However, any deviations from the original protocol should be supported by soundly based and well documented scientific arguments.

Concern is sometimes expressed that with the addition of further testing conditions and extra tester strains to ensure comprehensive testing, the simplicity, speed and economy of the *Salmonella* test is lost. In order to reduce the costs and effort involved, testing could be undertaken in a sequential way (Zeiger *et al.*, 1985). For example, when performing a preliminary test to select dose-levels for the main assays the strain which is most likely to be sensitive to the chemical structure can be used. This will maximise the chance of detecting mutagenic activity at the earliest opportunity. In the absence of other indications this often will be strain TA100.

In routine testing, work can sometimes stop if a positive result is obtained; extra conditions may be introduced as seems necessary, and full-scale experiments with independent replication may be needed only as is sufficient to demonstrate unequivocal negative results (see Section 2.4.4).

2.3.3.5 Exposure period

In the plate-incorporation test the period of exposure of the bacteria to the test substance is fixed, since the test substance, bacteria and S9 are added within a few seconds of each other.

In the pre-incubation assay, 60 minutes' exposure in buffer at 37 °C prior to adding soft agar is recommended for the reasons given previously. At the present time the need for aeration (i.e. use of shaking cultures) is considered optional.

In the treat-and-plate method the optimum time of exposure must be determined experimentally.

In the fluctuation test the period of exposure is again fixed for the duration of the test (i.e. 72 hours).

Ashby (1988) has recently recommended that the testing of liquids in *in vitro* genotoxicity assays should proceed mindful of the requirements that the test organism be **adequately** exposed to the test agent. The hydrolytic stability of the test chemical, as well as its volatility, should be considered when choosing the correct exposure conditions (see Section 2.3.1.1 (c)).

2.3.3.6 Scoring methods

Many laboratories now employ automatic electronic colony counters in order to determine the results of plate-incorporation tests. Such devices should be regularly calibrated against a series of authentic, hand-counted plates encompassing a range of mutant colonies, from very low to very high counts and colonies of varying sizes. Staff should be aware of the various artefacts (particles, air-bubbles, condensation, dirt on the lenses and optical surfaces) which can cause the counter to produce spurious results.

A recent paper by Claxton *et al.* (1984) documents the levels of accuracy for hand- and automated-counting techniques, and provides a simple method for generating acetate calibration transparencies for use with most colony counters. It should be recognised that with counts greater than 1500 colonies per plate the calculated correction factors may not be adequate.

2.4 DATA PROCESSING AND PRESENTATION

2.4.1 Recording and storage of data

This is discussed in general in Chapter 1.

2.4.2 Appropriate statistical analysis

The objective of applying statistical analysis to bacterial mutagenicity data is to help in determining whether the results of any assay are positive or negative. In addition, statistical methods may be used to assess the reproducibility of experiments and to determine a quantitative measure of effect. The use of the twofold rule in this assay is now generally considered too conservative (Mahon *et al.*, 1989).

Quantifying data facilitates comparison of results between different experiments, between different strains (e.g. between DNA-repair deficient strains which are otherwise isogenic), between assays done in different laboratories, and between closely related chemicals. It also enables computation of parameters such as the slope of dose-response curves, or the concentration required to double the mutation-frequency, useful values when conducting comparative studies. Quantifying bacterial mutagenicity data, although extremely valuable within the context outlined above, should not, however, be confused with 'external' uses of quantitative data, for example in numerological exercises claiming to show a useful association between mutagenic and carcinogenic potency (e.g. Meselson & Russell, 1977). Such claims have not been supported by any substantial evidence (Ashby & Styles, 1978).

The UKEMS Guidelines for Mutagenicity Testing Part III, Chapter 2 entitled 'Analysis of data from microbial colony assays' (Mahon *et al.*, 1989) and Chapter 4, entitled 'Statistical evaluation of bacterial/mammalian fluctuation tests' (Robinson *et al.*, 1989), specifically address the problem of carrying out valid statistical analysis of data generated from bacterial mutation assays (plate-incorporation assays, treat-and-plate tests, and fluctuation assays). Consequently, a detailed discussion of this subject is not included in these guidelines, and the reader is urged to consult the original document. The following major points were made:

1. As many microbial cells as is practical should be exposed to the test agent, and the results should clearly reflect this.
2. Protocols should be designed to limit the number of pre-existing mutants in starting cultures, and the validity of the test should be confirmed by checking that positive and negative control values fall within the historical control ranges.
3. Three methods of analysis, linear regression, Dunnett's method and Wahrendorf's method, can be recommended for the analysis of data from microbial colony assays; although each has its strengths and weaknesses.
4. The use of twice as many control plates will lead to more powerful tests and is recommended.
5. Experiments should be repeated, omitting any concentrations shown to be bactericidal in the first assay. When two experiments fail to agree, a third 'decider' experiment should be performed.
6. Statistical evaluation of the positive control is not necessary. Even if a concentration is used which is at or around the limit of sensitivity, a biologically significant result should be apparent.

Whatever test is chosen, it should be emphasised that the adoption of a statistical approach to analysing bacterial mutation data does not absolve the investigator from applying scientific acumen and common sense to the analysis of the biological significance of the data. Moreover, application of even the most arcane statistical analysis should not be expected to rescue poor data from badly conducted experiments.

2.4.3 Presentation of results: minimum data to be presented

The general requirements have been discussed in Chapter 1.

Presentation of raw data allows independent statistical evaluation. Where plate-tests have been performed, individual values for number of mutant colonies per plate should be tabulated: for fluctuation tests, numbers of positive tubes or wells should be given. Data from treat-and-plate experiments should include all the individual plate counts from dilution assays for viability as well as the raw data for number of mutant colonies per plate. Some regulatory authorities may require that results should be tabulated in ascending order of dose, starting with the solvent controls. Data from positive controls should be clearly identified as such and separated from the results obtained for the substance under test. The doses of test compound should be expressed by weight (per plate, or per ml) and not by volume although the actual volume added to the test system should be recorded. If the test substance is a formulation or mixture, results should also be expressed per weight of active ingredient(s).

In addition to the complete raw data set, if there is any evidence of activity, it may be helpful to display the results graphically to give a visual impression of the form of any dose response and the pattern of variability.

2.4.4 Interpretation of data in terms of positive and negative

For a substance to be considered positive in a plate-incorporation test and/or pre-incubation assay it should have induced a statistically significant dose-related increase in revertant count compared with appropriate concurrent controls in one or more strains of bacteria, in the presence and/or absence of S9, in at least two separate experiments. The problem of equivocal effects is discussed in Section 2.4.5.

A negative result is very much more difficult to define than a positive result, and such a definition must take into account the limits to which testing must be taken before deciding that the test substance is without mutagenic effect.

The Working Group defined a negative result as follows: A test substance can be considered negative if it produced no statistically significant increase in revertant count at any concentration according to the baseline protocol outlined in Fig. 2.1. This protocol includes the following requirements:

1. The use of the following bacterial strains:
   S. typhimurium TA1535, TA98, TA100
   S. typhimurium TA1537 **or TA97a**
   S. typhimurium TA102 **or *E.coli* WP2 *uvrA* (pKM101) and *E. coli* WP2 (pKM101).**
2. Tests conducted with at least five concentrations spaced at intervals differing by $\leqslant$3-fold, and extending either to the limits imposed by toxicity, solubility or if the substance is very soluble to an upper limit of 5 mg per plate.
3. Adequate concurrent negative and positive controls (for both +S9-mix and −S9-mix conditions of exposure).
4. Appropriate statistical analysis as defined by the UKEMS Statistical Guidelines (Mahon *et al.*, 1989).

2.4.5 Ambiguous results

In the context of bacterial mutation tests, 'ambiguous results' can be interpreted to include results where a positive outcome cannot be ruled out because at one or more concentrations there are significantly more revertants per plate than are seen on concurrent control plates, and that this increase is consistent in two or more experiments. The effect may or may not be dose-related, and might occur perhaps in just one tester strain and at one particular level of S9 in the S9-mix. Such a result cannot be classified as negative, neither is it positive. The use of historical control values to interpret ambiguous results is not recommended (Mahon *et al.*, 1989).

Ambiguous results may be caused by a technical problem, such as the presence of nutrients in the test substance or the bacteriostatic effect of the test substance; on the other hand, it might be an indication that a change in experimental procedure is required, as discussed in Section 2.3.3. In addition, in the course of several replicate experiments, one or two assays might be positive, and some might be negative. Results of this type may be classified as 'irreproducible'. Under these circumstances the use of alternative protocols may resolve the problem. If changes to the base-line protocol produce consistently positive results as defined in Section 2.4.4 the problem is resolved, otherwise the results remain ambiguous or irreproducible. This should be stated clearly, and the test substance should then be subjected to other types of short-term mutagenicity assay.

2.5 DISCUSSION

2.5.1 Influence of critical factors on the validity of the data

The conduct of bacterial mutation tests using well established protocols requires close attention to every aspect of the experimental procedure. Success in running large numbers of such tests in routine screening programmes depends on the establishment of consistent methods for every phase of the experiment. A deficiency in just one area will jeopardise the whole enterprise: once again the Working Group was unable to single out any particular factor as being critical in the conduct and interpretation of these tests. To identify particular points might be taken to mean that other factors were of lesser importance.

2.5.2 Interpretation of the results in terms of intrinsic mutagenic activity of the test material

A bacterial mutagenicity assay simply determines whether the substance under investigation is or is not a bacterial mutagen in the presence and/or absence of an exogenous metabolising system derived from a mammal. Using fluctuation tests or treat-and-plate tests, where mutagencity is determined at non-lethal doses, or where corrections are made for survival, it is possible to calculate mutation rates for the given bacterial species and strain in terms of number of mutations per bacterium or per cell division. It is a peculiarity of the plate-incorporation test that such absolute rates cannot be determined, since a quantitative measurement of the bacterial population, and hence toxicity, is not part of the standard assay nor the pre-incubation modification.

The result of a bacterial mutation test cannot of itself determine whether the test substance possesses genotoxic activity in any other species. Moreover, the potency of the test substance as a bacterial mutagen cannot usefully serve as a guide to its potency as a mutagen in higher organisms, especially mammals, which possess a multiplicity of organ systems whose function and activity can drastically modify or even abolish the mutagenic activity of a substance or its metabolite even before the target cells are reached. The implications of mutagenicity for carcinogenesis have already been discussed briefly in Section 2.1.2. There is little to suggest that there is any quantitative relationship between carcinogenicity and bacterial mutagenicity (see, for example, Coombs *et al.*, 1976; Ashby & Styles, 1978; Glatt *et al.*, 1979; Bartsch *et al.*, 1980).

Therefore the bacterial mutagenicity of a chemical should be regarded as a qualitative indication of hazard which should be weighed against any relevant data available for the test substance (e.g. *in vivo* test results, information on absorption, distribution, metabolism, excretion, proposed use, etc.).

2.6 CONCLUSIONS

Properly conducted bacterial mutation tests offer a convenient and well validated means of detecting DNA-damaging agents. It was apparent that the deleterious effects of a lapse in any one of several key areas (selection of appropriate solvent, choice and spacing of test concentrations, selection and maintenance of strains, preparation and use of metabolising systems) could not be mitigated by excellence in all of the others. Difficulties are still caused by ambiguous results, and it is recommended that scientifically sound deviations in base-line protocol should be encouraged in order to resolve or alleviate such problems.

2.7 REFERENCES

Aeschbacher, H.U. (1980). Mutagenicity testing of whole food products. In *Progress in Environmental Mutagenesis, Vol. 7*, ed. M. Alecevic. Elsevier, Amsterdam, pp. 201–5.

Aeschbacher, H.U., Finot, P.A. & Wolleb, U. (1983). Interactions of histidine containing test substances and extraction methods with the Ames mutagenicity test. *Mutation Research*, **113**, 103–16.

Albertini, S. & Gocke, E. (1988). Plasmid copy number and mutant frequencies in *S. typhimurium* TA102. *Environmental and Molecular Mutagenesis*, **12**, 353–63.

Alldrick, A.J. & Rowland, I.R. (1985). Activation of mutagens IQ and MeIQ by hepatic S9 fractions derived from various species. *Mutation Research*, **144**, 59–62.

Ames, B.N., McCann, J. & Yamasaki, E. (1975). Methods for detecting carcinogens and mutagens with the *Salmonella*/mammalian microsome mutagenicity test. *Mutation Research*, **31**, 347–64.

Anders, M.K., Karpinsky, G.E., McCoy, E.C. & Rosenkranz, H.S. (1982). Phenotypic instability of *Salmonella typhimurium* tester strains: An example of plasmid-enhanced genetic drift? *Mutation Research*, **97**, 411–28.

Anderson, D. & McGregor, D.B. (1980). The effect of solvents upon the yield of revertants in the *Salmonella*/activation mutagenicity assay. *Carcinogenesis*, **1**, 363–6.

Ashby, J. (1986). The prospects for a simplified and internationally harmonised approach to the detection of possible human carcinogens and mutagens. *Mutagenesis*, **1**, 3–17.

Ashby, J. (1988). The evaluation of volatile chemicals for mutagenicity. *Mutagenesis*, **4**, 160–2.

Ashby, J. & Styles, J.A. (1978). Does carcinogenic potency correlate with mutagenic potency in the Ames assay? *Nature*, **271**, 452–5.

Ashby, J. & Tennant, R.W. (1988). Chemical structure, *Salmonella* mutagenicity and extent of carcinogenicity as indicators of genotoxic carcinogenesis among 222 chemicals tested in rodents by the U.S. NCI/NTP. *Mutation Research*, **204**, 17–115.

Ashby, J., Callander, R.D. & Rose, F.L. (1985). Weak mutagenicity to *Salmonella* of the formaldehyde-releasing anti-tumour agent hexamethylmelamine. *Mutation Research*, **142**, 121–5.

Barber, E.D., Donish, W.H. & Mueller, K.R. (1983). The relationship between growth and reversion in Ames *Salmonella* plate incorporation assay. *Mutation Research*, **113**, 89–101.

Barnes, W.M., Tuley, E. & Eisenstadt, E. (1982). Base sequence analysis of his $^+$ revertants of the *his G46* missense mutation in *Salmonella typhimurium*. *Environmental Mutagenesis*, **4**, 297.

Bartsch, H., Camus, A.-M. & Malaveille, C. (1976). Comparative mutagenicity of N-nitrosamines in a semi-solid and in a liquid incubation system in the presence of rat or human tissue fractions. *Mutation Research*, **37**, 149–62.

Bartsch, H., Malaveille, C., Camus, A.-M., Martel-Planche, G., Brun, G., Hautefeuille, A., Sabadie, N., Barbin, A., Kuroki, T., Drevon, C., Piccoli, C. & Montesano, R. (1980). Bacterial and mammalian mutagenicity tests: validation and comparative studies on 180 chemicals, In *Molecular and Cellular Aspects of Carcinogen Screening Tests*, *IARC Scientific Publications No. 27*, ed. R. Montesano, H. Bartsch and L. Tomatis. International Agency for Research on Cancer, Lyon, pp. 179–241.

Bonneau, D. & Cordier, A. (1985). The susceptibility and discrimination of five *Salmonella typhimurium* strains in the Ames Test for routine screening. Abstract of the Fourth International Conference of Environmental Mutagens, Stockholm, p. 195.

Booth, S.C., Welch, A.M. & Garner, R.C. (1980). Some factors influencing mutant numbers in the *Salmonella*/microsome assay. *Carcinogenesis*, **1**, 911–23.

Bos, R.P., Neis, J.M., van Gemert, P.J.L. & Henderson, P.Th. (1983). Mutagenicity testing with the *Salmonella*/hepatocyte and *Salmonella*/microsome assays. A comparative study with some known genotoxic compounds. *Mutation Research*, **124**, 103–12.

Brennan-Craddock, W.E., Rowland, I.R., Mallett, A.K. & Neale, S. (1987). Age-dependent changes in activation of dietary mutagens by mouse hepatic fractions. *Mutagenesis*, **2**, 301–2.

Bridges, B.A. (1978). On the detection of volatile liquid mutagens with bacteria: experiments with dichlorvos and epichlorhydrin. *Mutation Research*, **54**, 367–71.

Bridges, B.A. & Mendelsohn, M.L. (1986). Recommendations for screening for potential human germ cell mutagens: An ICPEMC Working Paper No. I. In *Genetic Toxicology of Environmental Chemicals, Part B: Genetic Effects and Applied Mutagenesis*, ed. C. Ramel, B. Lambert and J. Magnusson. Liss, New York, pp. 51–65.

Bridges, B.A., Zeiger, E. & McGregor, D.B. (1981). Summary report on the performance of bacterial mutation assays. In *Evaluation of Short-Term Tests for Carcinogens, Progress in Mutation Research, Vol. 1*, ed. F.J. de Serres and J. Ashby. Elsevier, New York, pp. 49–67.

Brown, J.P. & Brown, R.J. (1976). Mutagenesis by 9,10-anthraquinone derivatives and related compounds in *S. typhimurium*. *Mutation Research*, **40**, 203–24.

Burnett, C., Fuchs, C., Corbett, J. & Menkart J. (1982). The effect of dimethylsulphoxide on the mutagenicity of the hair-dye, p-phenylene-diamine. *Mutation Research*, **103**, 1–4.

Busch, D.B., Archer, J., Amos, E.A., Hatcher, J.F. & Bryan, G.T. (1986). A protocol for the combined biochemical and serological identification of the Ames mutagen tester strains as *Salmonella typhimurium*. *Environmental Mutagenesis*, **8**, 741–51.

Callander, R.D. (1986). Observed convergence of the *Salmonella* plate and pre-incubation assays when employing varying levels of S9. *Mutagenesis*, **1**, 439–43.

Capon, D.J., Seeberg, P.H., McGrath, J.P., Hayflick, J.S., Edman, U., Levinson, A.D. & Goeddel, D.V. (1983). Activation of Ki-ras2 gene in human colon and lung carcinomas by two different point mutations. *Nature*, **304**, 507–13.

Carver, J.H., Machado, M.L. & MacGregor, J.A. (1985). Petroleum distillates suppress *in vitro* metabolic activation: higher [S9] required in the *Salmonella*/microsome mutagenicity assay. *Environmental Mutagenesis*, **7**, 369–80.

Cerniglia, C.E., Zhou, Z., Manning, B.W., Federle, T.W. & Heflich, R.H. (1986). Mutagenic activation of the benzidine-based dye Direct Black 38 by human intestinal microflora. *Mutation Research*, **175**, 11–16.

Chu, K.C., Patel, K.M., Lin, A.H., Tarone, R.E., Linhart, M.S. & Dunkel, V.C. (1981). Evaluating statistical analyses and reproducibility of microbial mutagenicity assays. *Mutation Research*, **85**, 119–32.

Claxton, L.D., Toney, S., Perry, E. & King, L. (1984). Assessing the effect of colony counting methods and genetic drift on Ames bioassay results. *Environmental Mutagenesis*, **6**, 331–42.

Cole, J., Arlett, C.F. & Green, M.H.L. (1976) The fluctuation test as a more sensitive system for determining induced mutation in L5178Y mouse lymphoma cells. *Mutation Research*, **41**, 377–86.

Combes, R., Anderson, D., Brooks, T., Neale, S. & Venitt, S. (1984). The detection of mutagens in urine, faeces and body fluids. In *UKEMS Sub-committee on Guidelines for Mutagenicity Testing. Report. Part II. Supplementary Tests*, ed. B.J. Dean. United Kingdom Environmental Mutagen Society, Swansea, pp. 203–44.

Coombs, M.M., Dixon, C. & Kissonerghis, A.-M. (1976). Evaluation of the mutagenicity of compounds of known carcinogenicity, belonging to the benz(a)anthracene, chrysene, and cyclopent(a)phenanthrene series, using Ames's test. *Cancer Research*, **3**, 4525–9.

Coulston, F. & Dunne, J.F. (1979). *The Potential Carcinogenicity of Nitrosatable Drugs, WHO Symposium*. Ablex Publishing Corporation, New Jersey, pp. 8–14.

De Flora, S. & Picciotto, A. (1980). Mutagenicity of cimetidine in nitrite enriched human gastric juice. *Carcinogenesis*, **1**, 925–30.

De Flora, S., Camoirano, S., Zanecchi, P. & Bennicelli, C. (1984a). Mutagenicity testing with TA97 and TA102 of 30 DNA-damaging compounds, negative with other *Salmonella* strains. *Mutation Research*, **134**, 159–65.

De Flora, S., Zanacchi, P., Camoirano, P., Bennicelli, P. & Badolati, G.S. (1984b). Genotoxic activity and potency of 135 compounds in the Ames reversion test and in a bacterial DNA-repair test. *Mutation Research*, **133**. 161–98.

De Serres, F.J. & Ashby, J. (1981). In *Evaluation of Short-term Tests for Carcinogens, Progress in Mutation Research, Vol. 1*, ed. F.J. de Serres and J. Ashby. Elsevier, New York, pp. 49–67.

De Serres, F.J. & Shelby, M.D. (1979). Recommendations on data production and analysis using the *Salmonella*/microsome mutagenicity assay. *Mutation Research*, **64**, 159–65.

Dunkel, V.C., Zeiger, E., Brusick, D., McCoy, E., McGregor, D., Mortelmans, K., Rosenkranz, H.S. & Simmon, V.F. (1984). Reproducibility of microbial mutagenicity assay: 1. Test with *Salmonella typhimurium* and *Escherichia coli* using a standardized protocol. *Environmental Mutagenesis*, **6** (Suppl. 2), 1–254.

Dyrby, T. & Ingvardsen, P. (1983). Sensitivity of different *E. coli* and *Salmonella* strains in mutagenicity testing calculated on the bases of selected literature. *Mutation Research*, **123**, 47–60.

EPA (1984). HG-Gene Muta-*S. typhimurium*, October 1984: The *Salmonella typhimurium* reverse mutation assay. Office of Toxic Subtances, Office of Pesticides and Toxic Substances, US Environmental Protection Agency, Washington DC 20460, USA.

Fasano, O., Aldrich, T., Tamanoi, F., Taparowsky, E., Farth, M. & Wigler, M. (1984). Analysis of the transforming potential of the human H-ras gene by random mutagenesis. *Proceedings of the National Academy of Sciences (USA)*, **81**, 4008–12.

Forster, R., Green, M.H. & Priestley, A. (1980). Optimal levels of S9 fraction in the Ames and fluctuation tests: apparent importance of diffusion of metabolites from top agar. *Carcinogenesis*, **1**, 337–46.

Forster, R., Green, M.H. & Priestley, A. (1981). Enhancement of S9 activation by S105 cytosolic fraction. *Carcinogenesis*, **2**, 1081–5.

Forster, R., Green, M.H.L., Gwilliam, R.D., Priestley, A. & Bridges, B.A. (1982). Use of the fluctuation test to detect mutagenic activity in unconcentrated samples of drinking waters in the United Kingdom. In *Water*

Chlorination: Environmental Impact and Health Effects, ed. R.L. Jolley *et al.* Ann Arbor Science, USA.
Gatehouse, D.G. (1978). Detection of mutagenic derivatives of cyclophosphamide and a variety of other mutagens in a Microtitre^R fluctuation test, without microsomal activation, *Mutation Research*, **53**, 289–96.
Gatehouse, D.G. & Delow, G.F. (1979). The development of a 'Microtitre^R' fluctuation test for detection of indirect mutagens, and its use in the evaluation of mixed enzyme induction of the liver. *Mutation Research*, **60**, 239–52.
Gatehouse, D.G. & Paes, D.J. (1983). A demonstration of the *in vitro* bacterial mutagenicity of procarbazine, using the microtitre fluctuation test and large concentrations of S9-fraction. *Carcinogenesis*, **4**, 347–52.
Gatehouse, D. & Wedd, D.J. (1984). The differential mutagenicity of isoniazid in fluctuation assays and *Salmonella* plate tests. *Carcinogenesis*, **5**, 391–7.
Gatehouse, D.G., Wedd, D. & Wharton, K. (1985). The comparative mutagenicity of 4-dimethylamino-azobenzene and 4-cyanodimethylaniline in plate incorporation tests and fluctuation assays. In *Comparative Genetic Toxicology*, ed. J.M. Parry and C.F. Arlett. MacMillan, Basingstoke, pp. 181–9.
Gibson, J.F., Boxer, P.G., Hedworth-Whitty, R.B. & Gompertz, D. (1983). Urine mutagenicity assays: a problem arising from the presence of histidine associated growth factors in XAD-2 prepared urine concentrates with particular relevance to assays carried out using the bacterial fluctuation test. *Carcinogenesis*, **4**, 1471–6.
Glatt, H.R., Schwind, H., Zajdela, F., Croisy, A., Jacquignon, P.C. & Oesch, F. (1979). Mutagenicity of 43 structurally related heterocyclic compounds and its relationship to their carcinogenicity. *Mutation Research*, **66**, 307–28.
Goggleman, W. (1980). Culture conditions and the influence of the number of bacteria on the number of spontaneous revertants per plate. In *Progress in Mutation Research, Vol. 2*, ed. A. Kappas. Elsevier/North Holland, Amsterdam, pp. 173–8.
Goggleman, W., Grafe, A., Vollmar, J., Baumeister, M., Kramer, P.J. & Pool, B.L. (1983). Criteria for standardisation of *Salmonella* mutagenicity tests. IV. Relationship between the number of his bacteria plated and the number of his$^+$ revertants scored in the test. *Teratogenesis, Carcinogenesis, Mutagenesis*, **3**, 205–13.
Goggleman, W. & Vollmar, J. (1985). Results of collaborative study with the new *Salmonella* strain TA102 (Abstract). *Mutation Research*, **147**, 132.
Grafe, A. & Goggleman, W. (1985). Studies with the new *Salmonella* strains TA97, TA102 and TA104 (Abstract). *Mutation Research*, **147**, 132–3.
Green, M.H.L. & Muriel, W.J. (1976). Mutagen testing using Tryp$^+$ reversion in *E. coli. Mutation Research*, **38**, 3–32.
Green, M.H.L., Muriel, W.J. & Bridges, B.A. (1976). Use of a simplified fluctuation test to detect low levels of mutagens. *Mutation Research*, **38**, 33–42.
Green, M.H.L., Bridges, B.A., Rogers, A.M., Horspool, G., Muriel, W.H., Bridges, J.W. & Fry, J.R. (1977a). Mutagen screening by a simplified bacterial fluctuation test: use of microsomal preparations and whole liver cells for metabolic activation. *Mutation Research*, **48**, 287–94.
Green, M.H.L., Rogers, A.M., Muriel, W.J., Ward, A.C. & McCalla, D.R. (1977b). Use of a simplified fluctuation test to detect and characterise mutagenesis by nitrofurans. *Mutation Research*, **44**, 139–43.

Hartman, P.E. (1983). Mutagens: some possible health impacts beyond carcinogenesis. *Environmental Mutagenesis*, **5**, 139–52.

Hartman, P.E. (1987). Identification of Ames mutagen tester strains as *Salmonella typhimurium*. *Environmental Mutagenesis*, **9**, 233–4.

Hartman, P.E., Ames, B.N., Roth, J.R., Barnes, W.N. & Levin, D.E. (1986). Target sequences for mutagenesis in *Salmonella* histidine-requiring mutants. *Environmental Mutagenesis*, **8**, 631–41.

Hartman, P.E. & Aukerman, S.L. (1986). *Salmonella* tester strains: mutational targets and correlation with animal carcinogenicity and teratogenicity. In *Mechanisms of DNA Damage and Repair*, ed. M.G. Simic, L. Grossman and A.D. Upton. Plenum Press, New York, pp. 407–24.

Hass, B.S., Heflich, R.H., Shaddock, J.G. & Casciano, D.A. (1986). Comparison of mutagenicities in a *Salmonella* reversion assay mediated by uninduced hepatocytes and hepatocytes from rats pretreated for 1 or 5 days with Aroclor 1254. *Environmental Mutagenesis*, **7**, 391–403.

Haworth, S., Lawlor, T., Mortelmans, K., Speck, W. & Zeiger, E. (1983). *Salmonella* mutagenicity results for 250 chemicals. *Environmental Mutagenesis* **5**, (Suppl. 1), 3–142.

Herbold, B.A. (1982). Preliminary data from an international survey on the sensitivity of Ames tester strains. *Mutation Research*, **97**, 191–2.

Hince, T.A. & Neale, S. (1977). Physiological modification of alkylating agent induced mutagenesis. I. Effect of growth rate and repair capacity on nitrosomethylurea-induced mutation of *E. coli*. *Mutation Research*, **46**, 1–10.

Hirota, H., Hayashi, K., Suzuki, Y. & Shimizu, H. (1987). The bubbling method for detecting mutagenic activity in gaseous compounds. *Mutation Research*, **182**, 359–60.

Hofnung, M. & Quillardet, P. (1986). Recent developments in bacterial short-term tests for the detection of genotoxic agents. *Mutagenesis*, **1**, 319–30.

Hubbard, S.A., Green, M.H.L., Gatehouse, D. & Bridges, J.W. (1984). The fluctuation test in bacteria. In *Handbook of Mutagenicity Test Procedures, 2nd edn*, ed. B.J. Kilbey, M. Legator, W. Nichols and C. Ramel. Elsevier, Amsterdam, pp. 142–61.

Hughes, T.J., Simmons, D.M., Monteith, L.G. & Claxton, L.D. (1987). Vaporization technique to measure mutagenic activity in volatile organic chemicals in the Ames/*Salmonella* Assay. *Environmental Mutagenesis*, **9**, 421–41.

Hyde, R., Smith, J.N. & Ioannides, C. (1987). Induction of the hepatic mixed-function oxidases by Arochlor 1254 in the hamster: comparison of Arochlor-induced rat and hamster preparations in the activation of pre-carcinogens in the Ames test. *Mutagenesis*, **2**, 477–82.

IARC (1980a). Report 2. Mutagenesis assays with bacteria. In *Long-term and Short-term Screening Assays for Carcinogens: a Critical Appraisal, IARC Monographs on the Evaluation of the Carcinogenic Risk of Chemicals to Humans, Supplment 2*. International Agency for Research on Cancer, Lyon, pp.85–106.

IARC (1980b). Report 10. Basic requirements for *in vitro* metabolic activation systems in mutagenesis testing, In *Long-term and Short-term Screening Assays for Carcinogens: a Critical Appraisal. IARC Monographs on the Evaluation of the Carcinogenic Risk of Chemicals to Humans, Supplement 2*. International Agency for Research on Cancer, Lyon, pp. 277–94.

ICPEMC (1983). Committee 1, Final Report. Screening strategy for chemicals that are potential germ cell mutagens in mammals, *Mutation Research*, **114**, 117–77.

Ishidate, M., Jr (1988). A proposed battery of tests for the initial evaluation of the mutagenic potential of medicinal and industrial chemicals. *Mutation Research*, **205**, 397–407.

JMHW (1984). *Guidelines for Testing of Drugs for Toxicity.* Pharmaceutials Affairs Bureau, Notice No. 118. Ministry of Health and Welfare, Japan.

Kado, N.Y., Langley, D. & Eisenstadt, E. (1983). A simple modification of the *Salmonella* liquid-incubation assay. Increased sensitivity for detecting mutagens in human urine. *Mutation Research*, **121**, 25–32.

Kappas A. (1988). On the mutagenic and recombinogenic activity of certain herbicides in *Salmonella typhimurium* and in *Aspergillus nidulans*. *Mutation Research*, **204**, 615–21.

Karpinsky, G.E. & Rosenkranz, H.S. (1980). The anaerobe-mediated mutagenicity of 2-nitrofluorene and 2-aminofluorene for *Salmonella typhimurium*. *Environmental Mutagenesis*, **2**, 253–58.

Kennelly, J.C., Stanton, C. & Martin, C.N. (1984). The effect of acetyl-CoA supplementation on the mutagenicity of benzidines in the Ames assay. *Mutation Research*, **137**, 39–45.

Kerklaan, P.R.M., Zoetemelk, C.E.M. & Mohn, G. (1985). Mutagenic activity of various chemicals in *Salmonella* strain TA100 and glutathione-deficient derivatives. *Biochemical Pharmacology*, **34**, 2151–6.

Kier, L.E., Brusick, D.J., Auletta, A.E., Von Halle, E.S., Brown, M.M., Simmon, V.F., Dunkel, V., McCann, J., Mortelmans, K., Prival, M., Rao, T.K. & Ray. V. (1986). The *Salmonella typhimurium*/mammalian microsomal assay. A report of the U.S. Environmental Protection Agency Gene-Tox Program. *Mutation Research*, **168**, 69–240.

Le, J., Jung, R. & Kramer, M. (1985). Effects of using liver fractions from different mammals, including man, on results of mutagenicity assay, in *Salmonella typhimurium*. *Food and Chemical Toxicology*, **23**, 695–700.

Lefevre, P.A. & Ashby, J. (1981). The effects of pre-incubation period and norharman on the mutagenic potency of 4-dimethylaminoazobenzene and 3-methyl-4-dimethylaminoazobenzene. *Carcinogenesis*, **2**, 927–31.

Levin, D.E. & Ames, B.N. (1986). Classifying mutagens as to their specificity in causing the six possible transitions and transversions: a simple analysis using the *Salmonella* mutagenicity assay. *Environmental Mutagenesis*, **8**, 9–28.

Levin, D.E., Hollstein, M., Christman, M.F., Schwiers, E.A. & Ames, B.N. (1982a). A new *Salmonella* tester strain (TA 102) with A–T base pairs at the site of mutation detects oxidative mutagens. *Proceedings of the National Academy of Sciences (USA)*, **79**, 7445–9.

Levin, D.E., Marnett, L.J. & Ames, B.N. (1984). Spontaneous and mutagen-induced deletions: Mechanistic studies in *Salmonella* tester strain TA 102. *Proceedings of the National Academy of Sciences (USA)*, **81**, 4457–61.

Levin, D.E., Yamasaki, E. & Ames, B.N. (1982b). A new *Salmonella* tester strain, TA 97, for the detection of frameshift mutagens. A run of cytosines as a mutational hot-spot. *Mutation Research*, **94**, 315–30.

Lijinsky, W. & Andrews, A.W. (1983). The superiority of hamster liver microsomal fraction for activating nitrosamines to mutagens in *Salmonella typhimurium*. *Mutation Research*, **111**, 135–44.

Lindblad, W. J. & Jackim, E. (1982). Mechanism for the differential induction of mutation by S9 activated benzo(a)pyrene employing either a Glucose-6-phosphate dependent NADPH-regenerating system or an Isocitrate dependent system. *Mutation Research*, **96**, 109–18.

Luria, S.E. & Delbruck, M. (1943). Mutations of bacteria from virus sensitivity to virus resistance. *Genetics*, **28**, 491–511.

Mahon, G.A.T., Green, M.H.L., Middleton, B., Mitchell, I. de G., Robinson, W.D. & Tweats, D.J. (1989). Analysis of data from microbial colony assays. In *UKEMS Sub-committee on Guidelines for Mutagenicity Testing. Report. Part III. Statistical Evaluation of Mutagenicity Test Data*, ed. D.J. Kirkland. Cambridge University Press, Cambridge, pp.28–65.

Margolin, B.H., Risko, K.J., Shelby, M.D. & Zeiger, E. (1983). Sources of variability in Ames *Salmonella typhimurium* tester strains: analysis of the International Collaborative Study on 'Genetic Drift'. *Mutation Research*, **130**, 11–25.

Marino, D.J. (1987). Evaluation of Pluronic Polyol F127 as a vehicle for petroleum hydrocarbons in the *Salmonella*/microsomal Assay. *Environmental Mutagenesis*, **9**, 307–16.

Maron, D.M. & Ames, B.N. (1983). Revised methods for the *Salmonella* mutagenicity test. *Mutation Research*, **113**, 173–215.

Maron, D., Katzenellenbogen, J. & Ames, B.N. (1981). Compatability of organic solvents with the *Salmonella*/microsome test. *Mutation Research*, **88**, 343–50.

Matsushima, T., Sawamura, M., Hara, K. & Sugimura, T. (1978). A safe substitute for polychlorinated biphenyls as an inducer of metabolic activation system. In *In vitro Metabolic Activation in Mutagenicity Testing*, ed. F.J. de Serres, J. Foutes, J.R. Bend and R. Philpot. Elsevier, North Holland, pp. 85–6.

McCann, J. & Ames, B.N. (1976). Detection of carcinogens as mutagens in the *Salmonella*/microsome test: assay of 300 chemicals: discussion. *Proceedings of the National Academy of Sciences (USA)*, **73**, 950–4.

McCann, J., Choi, E., Yamasaki, E. & Ames, B.N. (1975). Detection of carcinogens as mutagens in the *Salmonella*/microsome test: assay of 300 chemicals. *Proceedings of the National Academy of Sciences (USA)*, **72**, 5135–9.

McCoy, E.C., Speck., W.T. & Rosenkranz, H.S. (1977). Activation of a procarcinogen to a mutagen by a cell-free extract of anaerobic bacteria. *Mutation Research*, **46**, 261–4.

McGregor, D.B. & Prentice, R.D. (1985). Phenobarbital: its mutagenicity and toxicity in the Ames *Salmonella* test. In *Evaluation of Short-term Tests for Carcinogens. Progress in Mutation Research, Vol. 5*, ed. J. Ashby *et al.* Elsevier, Amsterdam–Oxford–New York, pp. 741–3.

McGregor, D., Prentice, R.D., McConville, M., Lee, Y.J. & Caspary, W.J. (1984). Reduced mutant yield at high doses in the *Salmonella*/activation assay: the cause is not always toxicity. *Environmental Mutagenesis*, **6**, 545–57.

MacGregor, J.T. & Wilson, R.E. (1985). Mutagenicity tests of lipid oxidation products in *Salmonella typhimurium*: monohydroperoxides and secondary oxidation products of methyl linoleate and ethyl linoleate. *Food and Chemical Toxicology*, **23**, 1041–7.

McMahon, R.E., Cline, J.C. & Thompson, C.Z. (1979). Assay of 855 test chemicals in ten tester strains using a new modification of the Ames test for bacterial mutagens. *Cancer Research*, **39**, 682–93.

McPhee, D.G. (1984). Do the *Salmonella typhimurium* tester strains used in mutagenicity assays display 'plasmid-enhanced genetic drift'? *Mutation Research*, **127**, 183–4.

Meselson, M. & Russell, K. (1977). Comparisons of carcinogenic and mutagenic potency, In *Origins of Human Cancer, Book C: Human Risk*

Assessment, ed. J.J. Hiatt, J.D. Watson and J.A. Winsten. Cold Spring Harbor Laboratory, New York, pp. 1473–81.

Miller, J.A. & Miller, E.C. (1976). The metabolic activation of chemical carcinogens to reactive electrophiles. In *Biology of Radiation Carcinogenesis*, ed. J.M. Yuhas, R.W. Tennant and J.D. Regan. Raven Press, New York, pp. 147–64.

Mitchell, I. de G., Dixon, P.A., Gilbert, P.J. & White, D.J. (1980). Mutagenicity of antibiotics in microbial assays. Problems of evaluation. *Mutation Research*, **79**, 91–105.

Mohn, G., Kerklaan, P. & Ellenberger, J. (1984). Methodologies for the direct and animal mediated determination of various genetic effects in derivatives of strain 343/113 of *E. coli* K.12. In *Handbook of Mutagenicity Test Procedures, 2nd edn*, ed. B.J. Kilbey, M. Legator, W. Nichols and C. Ramel. Elsevier, Amsterdam, pp. 189–214.

Moriya, M., Ohta, T., Watanabe, K., Miyazawa, T., Kato, K. & Shirasu, Y. (1983). Further mutagenicity studies on pesticides in bacterial reversion assay systems. *Mutation Research*, **116**, 185–216.

Mortelmans, K., Haworth, S., Lawlor, T., Speck, W., Tainer, B. & Zeiger, E. (1985). *Salmonella* mutagenicity tests: II. Results from the testing of 270 chemicals. *Environmental Mutagenesis*, **8** (Suppl. 7), 1–119.

Nestmann, E.R., Douglas, G.R., Kowbel, D.J. & Harrington, T.R. (1985). Solvent interactions with test compounds and recommendations for testing to avoid artifacts. *Environmental Mutagenesis*, **7**, 163–70.

Neudecker, T. & Henschler, D. (1985a). Allyl isothiocyanate is mutagenic in *Salmonella typhimurium*. *Mutation Research*, **156**, 33–7.

Neudecker, T. & Henschler, D. (1985b). Mutagenicity of chloro-olefins in the *Salmonella*/mammalian microsome test. I. Allyl chloride mutagenicity re-examined. *Mutation Research*, **157**, 145–8.

Nunoshiba, T. and Nishioka, H. (1984). Protective effect of R-Factor plasmid pKM101 on lethal damage by UV and chemical mutagens in *E. coli* strains with different DNA repairing capacities. *Mutation Research*, **141**, 135–9.

OECD (1983). *OECD Guidelines for the Testing of Chemicals, 471*. Organisation for Economic Cooperation and Development, Paris, 1983.

O'Neill, J.P., Brimer, P.A. & Hsie, A.H. (1981). Fluctuation analyses of spontaneous mutations to 6-thioguanine resistance in Chinese hamster ovary cells in culture. *Mutation Research*, **82**, 343–53.

Ong, T., Mukhtar, M., Wolf, C.R. & Zeiger, E. (1980). Differential effects of cytochrome P450-inducers on promutagen activation capabilities and enzymatic activities of S-9 from rat liver. *Journal of Environmental Pathology and Toxicology*, **4**, 55–65.

Paes, D.J. (1984). Letter to the Editor: Microbial mutagenicity of selected hydrazines: misuse of data. *Mutation Research*, **136**, 89–90.

Pagano, D.A. & Zeiger, E. (1985). The stability of mutagenic chemicals stored in solution. *Environmental Mutagenesis*, **7**, 293–302.

Parry, J.M. (1977). The use of yeast cultures for the detection of environmental mutagens using a fluctuation test. *Mutation Research*, **46**, 165–76.

Phillips, B.J., Kranz, E., Elias, P.S. & Munzner, R. (1980). An investigation of the genetic toxicology of irradiated foodstuffs using short term test systems. 1. Digestion *in vitro* and the testing of digests in the *Salmonella typhimurium* reverse mutation test. *Food and Cosmetics Toxicology*, **18**, 371–75.

Piper, C.E. & Kuzdas, C.D. (1987). Incorporation of TA97a into a standard Ames test protocol (abstract). *Environmental Mutagenesis*, **9** (Suppl.8), 85.

Prival, M.J., Bell, S.J., Mitchell, V.D., Peiperl, M.D. & Vaughn, V.L. (1984).

Mutagenicity of benzidine and benzidine-congener dyes and selected monoazo dyes in a modified *Salmonella* assay. *Mutation Research*, **136**, 33–47.

Prival, M.J., Davis, M., Peiperl, M. & Bell, S. (1986). Methods for detecting mutagens in certified azo colours used in foods. *Environmental Mutagenesis*, **8** (Suppl. 6), 66.

Prival, M.J. & Mitchell, V.D. (1982). Analysis of a method for testing azo-dyes for mutagenic activity in *S. typhimurium* in the presence of FMN and hamster liver S9. *Mutation Research*, **97**, 103–16.

Pueyo, C. & Hera, C. (1986). Conditions for optimal use of the L-arabinose-resistance mutagenesis test with *Salmonella typhimurium*. *Mutagenesis*, **1**, 267–74.

Purchase, I.F.H., Longstaff, E., Ashby J., Styles, J.A., Anderson, D., Lefevre, P.A. & Westwood, F.R. (1978). An evaluation of six short-term tests for detecting organic chemical carcinogens. *British Journal of Cancer*, **37**, 873–959.

Raineri, R., Andrew, A.W. & Poiley, J.A. (1986). Effect of donor age on the levels of activity of rat, hamster, and human liver S9 preparations in the *Salmonella* mutagenicity assay. *Journal of Applied Toxicology*, **6**, 101–8.

Reddy, E.P., Reynolds, R.K., Santos, E. & Barbacid, M. (1982). A point mutation responsible for the acquisition of transforming properties by the T24 human bladder carcinoma oncogene. *Nature*, **300**, 149–52.

Rinkus, S.J. & Legator, M.S. (1979). Chemical characterisation of 465 known or suspected carcinogens and their correlation with mutagenic activity in the *Salmonella typhimurium* system. *Cancer Research*, **39**, 3289–318.

Robertson, J., Harris, W.J. & McGregor, D.B. (1982a). Factors affecting the response of N,N-dimethylaminoazobenzene in the Ames microbial mutation assay. *Carcinogenesis*, **3**, 977–80.

Robertson, J.A., Harris, W.J. & McGregor, D.B. (1982b). Mutagenicity of azo-dyes in the *Salmonella*/activation test. *Carcinogenesis*, **3**, 21–5.

Robinson, W.D., Healy, M.J.R., Green, M.H.L., Cole, J., Gatehouse, D.G. & Garner, R.C. (1989). Statistical evaluation of bacterial/mammalian fluctuation tests. In *UKEMS Sub-committee on Guidelines for Mutagenicity Testing. Report. Part III. Statistical Evaluation of Mutagenicity Test Data*, ed. D.J. Kirkland. Cambridge University Press, Cambridge, pp.102–40.

Rosenkranz, E.J., McCoy, E.C., Mermelstein, R. & Rosenkranz, H.S. (1982). Evidence for the existence of distinct nitroreductases in *S. typhimurium*: roles in mutagenesis. *Carcinogenesis*, **3**, 121–3.

Rowland, I., Rubery, E. & Walker, R. (1984). Bacterial assays for mutagens in food. *UKEMS Sub-committee on Guidelines for Mutagenicity Testing. Report. Part II. Supplementary Tests*, ed. B.J. Dean. United Kingdom Environmental Mutagen Society, Swansea, pp. 173–202.

Rumruen, K. & Pool, B.L. (1984). Metabolic activation capabilities of S9 and hepatocytes from uninduced rats to convert carcinogenic N-nitrosamines to mutagens. *Mutation Research*, **140**, 147–53.

Ryan, F.J. (1955). Spontaneous mutation in non-dividing bacteria. *Genetics*, **40**, 726–38.

Sakai, M., Yoshida, D. & Mizusaki, S. (1985). Mutagenicity of polycyclic aromatic hydrocarbons and quinones on *Salmonella typhimurium* TA97. *Mutation Research*, **156**, 61–7.

Salmeen, I. & Durisin, A.M. (1981). Some effects of bacterial population on quantitation of Ames *Salmonella*-histidine reversion mutagenesis assays. *Mutation Research*, **85**, 109–18.

Scribner, H.E., McCarthy, K.L., Moss, J.M., Hayes, A.W., Smith, J.M., Cifone, M.A., Probit, G.S. & Valencia, R. (1983). The genetic toxicology of Kathon biocide, a mixture of 5-chloro-2-methyl-4-isothiazolin-3-one and 2-methyl-4-isothiazolin-3-one. *Mutation Research*, **118**, 129–52.

Shahin, M.M., Chopy, C. & Lequesne, N. (1985). Comparison of mutation induction by six monocyclic aromatic amines in *Salmonella typhimurium* tester strains TA97, TA1537, and TA1538. *Environmental Mutagenesis*, **7**, 535–46.

Sims, P. (1980). Metabolic activation of chemical carcinogens. *British Medical Bulletin*, **36**, 11–18.

Skopek, T.R., Liber, H.L., Krolewski, J.J. & Thilly, W.G. (1978). Quantitative forward mutation assay in *Salmonella typhimurium*, using 8-azaguanine resistance as a genetic marker. *Proceedings of the National Academy of Sciences (USA)*, **75**, 410–14.

Solt, A.K. & Neale, S. (1979). Induction of bacterial mutants in rodents treated with N-nitroso compounds. *Mutation Research*, **64**, 147.

Speck, W.T., Ellner, P.D. & Rosenkranz, H.S. (1975). Mutagenicity testing with *S. typhimurium* tester strains. I. Unusual phenotypes of the tester strains. *Mutation Research*, **28**, 27–30.

Suter, W. & Jaeger, I. (1981). Aroclor 1254 and 5,6-benzoflavone/phenobarbital-induced rat liver homogenate: comparison of enzymatic activities and metabolic potency. *Mutation Research*, **85**, 261–2.

Tabin, C.J., Bradley, S.M., Bargmann, C.I. & Weinberg, R.A. (1982). Mechanism of activation of a human oncogene. *Nature*, **300**, 143–9.

Tamura, G., Gold, C., Ferro-Luzzi, A. & Ames, B.N. (1980). Fecalase; a model for activation of dietary glycosides to mutagens by intestinal flora. *Proceedings of the National Academy of Sciences (USA)*, **77**, 4961–5.

Tennant, R.W., Margolin, B.H., Shelby, M.D., Zeiger, E., Haseman, J.K., Spalding, J., Caspary, W., Resnick, M., Staseiwicz, S., Anderson, B. & Minor, R. (1987). Prediction of chemical carcinogenicity in rodents from *in vitro* genetic toxicity assays. *Science*, **236**, 933–41.

Thomas, H.F. & Cole, J.A. (1986). Effects of excision repair and plasmid pKM101 on mutagenic and cytotoxic potencies of anthracycline derivatives in test strains of *S. typhimurium*. *Environmental Mutagenesis*, **8**, 797–815.

Thomas, S.M. & McPhee, D.G. (1984). A DNA-repair proficient strain of *Escherichia coli* which is highly sensitive to mutagenic acridines in plate tests. *Mutation Research*, **131**, 193–6.

Tosk, J., Schmeltz, I. & Hoffman, D. (1979). Hydrazines as mutagens in a histidine requiring auxotroph of *S. typhimurium*. *Mutation Research*, **66**, 247–52.

Venitt, S. (1978). Letter to the Editor, *Mutation Research*, **57**, 107–13.

Venitt, S. (1982). UKEMS Collaborative Genotoxicity Trial: Bacterial mutation tests of 4-chloromethylbiphenyl, 4-hydroxymethylbiphenyl and benzyl chloride: analysis of data from seventeen laboratories, *Mutation Research*, **100**, 91–109.

Venitt, S. & Bosworth, D. (1983). The development of anaerobic methods for bacterial mutation assays: aerobic and anaerobic fluctuation tests of human faecal extracts and reference mutagens. *Carcinogenesis*, **4**, 339–45.

Venitt, S. & Crofton-Sleigh, C. (1981). Mutagenicity of 42 coded compounds in a bacterial assay using *Escherichia coli* and *Salmonella typhimurium*. In *Evaluation of Short-term Tests for Carcinogens. Report of the International Collaborative Program. Progress in Mutation Research, Vol. 1*, ed. F.J. de Serres and J. Ashby. Elsevier, New York, pp. 351–60.

Venitt, S., Crofton-Sleigh, C. & Forster, R. (1984). Bacterial mutation assays using reverse mutation. In *Mutagenicity Testing: A Practical Approach*, ed. S. Venitt and J.M. Parry. IRL Press, Oxford, pp. 45–97.

Venitt, S., Forster, R. & Longstaff, E. (1983). Bacterial Mutation Tests. In *UKEMS Sub-committee on Guidelines for Mutagenicity Testing. Report. Part I. Basic Test Battery*, ed. B.J. Dean. United Kingdom Environmental Mutagen Society, Swansea, pp. 5–40.

Vithayathil, A.J., McClure, C. & Myers, J.W. (1983). *Salmonella*/microsome multiple indicator mutagenicity test. *Mutation Research*, **121**, 33–7.

Voogd, C.E., Van der Stel, J.J. & Jacobs, J.J.J.A.A. (1974). The mutagenic action of nitroimidazoles. I. Metronidazole, nimorazole, dimetridazole and ronidazole. *Mutation Research*, **26**, 483–90.

Waleh, N.S., Rapport, S.J. & Mortelmans, K. (1982). Development of a toxicity test to be coupled to the Ames *Salmonella* assay and the method of construction of the required strains. *Mutation Research*, **97**, 247–56.

Wedd, D.J., Burke, D. & Wilcox, P. (1988). Investigation into factors affecting the duration of lag-phase in the Ames test. *Mutagenesis*, **3**, 445–6.

Wilcox, P., Naidoo, A., Wedd, D.J. & Gatehouse, D.G. (1990). Comparison of *Salmonella typhimurium* TA102 with *Escherichie coli* WP2 tester strains. *Mutagenesis* **5**, 285–91.

Wolff, S. (1977). *In vitro* inhibition of mono-oxygenase dependent reactions by organic solvents. International Conference on Industrial and Environmental Xenobiotics. Prague.

Yahagi, T., Degawa, M., Seino, Y., Matsushima, T., Nagao, M., Sugimura, T. & Hashimoto, Y. (1975). Mutagenicity of carcinogen azo dyes and their derivatives. *Cancer Letters*, **1**, 91–6.

Yahagi, T., Nagao, M., Seino, Y., Matsushima, T., Sugimura, T. & Okado, M. (1977). Mutagenicities of N-nitrosamines on *Salmonella*. *Mutation Research*, **48**, 121–30.

Yamanaka, H., Nagao, M., Sugimura, T., Furuya, T., Shirai, A. & Matsushima, T. (1979). Mutagenicity of pyrrolozidine alkaloids in the *Salmonella*/mammalian-microsome test. *Mutation Research*, **68**, 211–6.

Zeiger, E. (1987). Carcinogenicity of mutagens: predictive capability of the *Salmonella* mutagenesis assay for rodent carcinogenicity. *Cancer Research*, **47**, 1287–96.

Zeiger, E. Anderson, B., Haworth, S., Lawlor, T. & Mortelmans, K. (1988). *Salmonella* mutagenicity tests: IV. Results from testing of 300 chemicals. *Environmental and Molecular Mutagenesis*, **11** (Suppl. 12), 1–158.

Zeiger, E. & Haworth, S. (1985). Tests with a preincubation modification of the *Salmonella*/microsome assay. In *Evaluation of Short-term Tests for Carcinogens. Progress in Mutation Research, Vol. 5*. ed. J. Ashby *et al.* Elsevier, Amsterdam–Oxford–New York, pp. 187–99.

Zeiger, E. & Pagano, D.A. (1984). Suppressive effects of chemicals in mixture on the *Salmonella* plate test response in the absence of apparent toxicity. *Environmental Mutagenesis*, **6**, 683–94.

Zeiger, E., Risko, K.J. & Margolin, B.H. (1985). Strategies to reduce the cost of mutagenicity screening with the *Salmonella* Assay. *Environmental Mutagenesis*, **7**, 901–11.

3

Metaphase chromosome aberration assays *in vitro*

D. SCOTT B.J. DEAN
N.D. DANFORD D.J. KIRLAND

3.1 INTRODUCTION

Cytogenetic tests are based upon the detection of certain chromosome changes using the light microscope. These changes are: structural chromosome aberrations (CA), sister chromatid exchanges (SCE) and numerical changes. Assays for SCE are dealt with in the UKEMS Report, Part II (Perry *et al.*, 1984). This Report deals with structural chromosome aberrations.

Chromosome aberration assays aim to detect the induction of chromosome breakage (clastogenesis). Although chemicals produce structural chromosome aberrations by a variety of mechanisms, the end-point is a discontinuity in the chromosomal DNA which is either rejoined or repaired to restore the original structure, left unrejoined, or rejoined inaccurately to produce a chromosome rearrangement (Bender *et al.*, 1974). Gross, readily visible CA such as terminal deletions and exchanges (Table 3.1) are recorded. Many of these will be lethal to the cell during the first few cell cycles after their induction but are used as indicators of the presence of non-lethal changes such as reciprocal translocations, inversions and small deletions. These more subtle changes may have important consequences in both germ and somatic cells.

The presence of structural or numerical aberrations in germ cells can lead to dominant lethality, perinatal mortality or congenital malformations in the survivors (Chandley, 1981; Sankaranarayanan, 1982). In somatic cells they may play a part in processes leading to malignancy. A better understanding of the role of CA in neoplasia has come from studies of potential oncogenes (proto-oncogenes). If translocations occur at the sites of these genes their expression can be changed as a consequence of their re-location (Minden, 1987). Chromosome deletions and rearrangements or total chromosome loss can lead to the elimination of tumour suppressor genes, resulting in malignancy (Phillips, 1987).

Chromosome aberration tests *in vitro* usually utilise mammalian somatic cells, the most popular being human peripheral blood lymphocytes or Chinese hamster fibroblasts in long-term culture. A disadvantage of many cultured mammalian cells is their limited ability to activate metabolically some potential clastogens. This can be overcome by adding an exogenous metabolic activation system, such as S9-mix, to the cells (Natarajan *et al.*, 1976; Madle & Obe, 1980).

Observations are made in metaphase cells arrested with a spindle inhibitor such as Colcemid, before hypotonic treatment, fixation and staining.

There are two significant changes to these current guidelines compared with the 1983 version (Scott *et al.*, 1983). The first is the recommendation that all tests be repeated regardless of the outcome of the first test. The second is that if a negative or equivocal result is obtained in the first test, the repeat test should include an additional sampling time as indicated below.

Recommendations for sampling time in the 1983 UKEMS Guidelines were based on available data in which cells had been harvested at various times after clastogen treatment of asynchronous populations and on a knowledge of cell cycle kinetics. The recommendation was for a single sample at approximately 1.5 normal cycle times from the beginning of treatment, provided that a range of concentrations was used which induced marginal to substantial reductions in mitotic index, which is generally indicative of mitotic delay. However, Ishidate (1988a) has since reported a significant number of chemicals which gave negative responses with a fixation time of 24 hours but were positive at 48 hours, using a Chinese hamster fibroblast line (CHL) with a doubling time of 15 hours. These observations suggest that there may be chemicals which induce very extensive mitotic delay at clastogenic doses or may be clastogenic only when cells have passed through more than one cell cycle since the beginning of treatment (see Thust *et al.*, 1980).

We now recommend, therefore, that if negative or equivocal results are obtained with a single harvesting time at 1.5 normal cycle times, the repeat test should include an additional sample at approximately 24 hours later. It may be necessary only to score cells from the top dose at this later fixation time (see Section 3.3.2.5). If the first test gives a clearly positive result the repeat test need utilise only the same fixation time. The use of more than one sampling time brings us into line with other Guidelines (e.g. European Communities, 1984; Japanese Guidelines: see Ishidate, 1988b; American Society of Testing and Materials: see Preston *et al.*, 1987). The need for the additional sampling time in repeat tests using our revised protocol should be reviewed when a data-base of sufficient size has been accumulated.

Table 3.1. *Structural chromosome aberrations: classification and definitions (based upon ISCN, 1985, and Savage, 1976, 1983)*

Aberration type	Diagrams with symbols[3]	Definitions and comment
A. Chromosome type (*cs*)		Involving both chromatids of a chromosome at identical loci
(1) Chromosome gap	(*csg*,G)	A non-staining region or achromatic lesion at the same locus in both chromatids with minimal misalignment of the chromatids[1]
(2) Chromosome break	(*ace*)	A discontinuity at the same locus in both chromatids giving an acentric fragment (*ace*). The abnormal (shortened) monocentric chromosome may not be identified.[4] Where sister union occurs the aberration should be classified as a chromatid-type 'isochromatid break'[2]
(3) Chromosome exchange (*cse*)		Involving two or more loci in the same or different chromosome(s)
(a) Interchange, between chromosomes (C/C)	e.g. Dicentric with associated fragment (*dic* + *ace*)	The dicentric results from an asymmetrical exchange that also produces a fragment which should not be scored as a separate event. Symmetrical exchanges that produce neither dicentrics nor fragments are not usually scored because of difficulties in detection

Table 3.1. *Contd.*

Aberration type	Diagrams with symbols[3]	Definitions and comment
(b) Intrachange, within a chromosome (C͡/C)		
(i) between arms (inter-arm intrachange)	 e.g. Centric ring with fragment (*csr* + *ace*)	The fragment is part of the exchange and should not be scored as a separate event. One fragment should be allocated to each ring (or dicentric)
(ii) within an arm (intra-arm intrachange)	 (*cs min*, M)	The centric part of this exchange may not be identified. Acentric rings are formed and are generally small and called minutes (*cs min* or M) or interstitial deletions[5]
B. Chromatid aberrations (*ct*)		Usually involve only one chromatid of a chromosome except for isochromatid breaks[2]
(1) Chromatid gap	(*ctg*, g)	A non-staining region or achromatic lesion in which there is minimal misalignment of the chromatid[1]
(2) Chromatid break	(*ctb*, c)	A discontinuity in which there is a clear misalignment of one of the chromatids[1]

Table 3.1. *Contd.*

Aberration type	Diagrams with symbols[3]	Definitions and comment
(3) Isochromatid break (i)[2,5]	 (SU)	Showing complete rejoining or sister union (SU) of broken ends
	(Nud) (Nup)	Incomplete rejoining (non-union, Nu) either proximally (p) or distally (d). Fragments may be aligned or displaced
(4) Chromatid exchange (*cte*)		
(a) Interchange, between chromosomes (c/c)		
(i) Asymmetrical	 (*qr*, c/c)	An acentric fragment and a dicentric chromatid are produced if rejoining is complete. Sometimes called a quadriradial (*qr*)
(ii) Symmetrical	 (*qr*, c/c)	Does not lead to a dicentric chromatid or to an acentric fragment unless the rejoining is incomplete. Also called a quadriradial (*qr*)
(b) Intrachange, within a chromosome ($\widehat{c/c}$)		
(i) between arms (inter-arm intrachange)	 Asymmetrical (*ctr*)	Centric ring formed
	 Symmetrical (*ct inv*)	Produces an inversion (*inv*) in a chromatid

Table 3.1. *Contd.*

Aberration type	Diagrams with symbols[3]	Definitions and comment
(ii) within an arm (intra-arm intrachange)	Minute (*ct min*, m) or interstitial deletion[5]	Often remains associated with the chromosome of origin
(b) Isochromatid/chromatid interchange (i/c) giving a triradial (*tr*)	 Dicentric triradial (*tr (2 cen)*, i/c)	Fragment produced and is not an independent aberration
	 Monocentric triradial (*tr (1 cen)*, i/c)	This should be counted as just one aberration as for all exchanges (see text, Section 4.1)

The following changes have been made from the 1983 UKEMS Guidelines:

1. The distinction between gaps and breaks is based upon the alignment of acentric fragments rather than on the size of the non-staining region, to comply with ISCN 1985.
2. The category of 'isolocus breaks' has been removed. Double acentric fragments should either be called acentric chromosome fragments (*ace*) or isochromatid breaks (chromatid-type aberrations) if there is proximal or distal sister union.
3. ISCN abbreviations are italicised, those used in the 1983 UKEMS Guidelines are not.
4. The symbol *csb* (chromosome break) should only be used when the origin of the *ace* as a simple terminal deletion is established; this would normally require chromosome banding.
5. It is often difficult to distinguish between the various intra-arm events [e.g. acentric fragments (ace), minutes (min) and isochromatid breaks] so for convenience these are listed as deletions although they may be derived from exchange events. Inter-arm and inter-chromosome events are listed under exchanges.

This table indicates the main types of chromosome aberrations which are scored. A more detailed treatment can be found in Savage (1976, 1983). We are most grateful to Dr J. R. K. Savage for providing the chromosome drawings.

3.2 THE TEST MATERIAL

General recommendations relating to test materials are given in Chapter 1 of these Guidelines.

Cultured cells are normally treated in their usual growth medium, so the solubility of the test material in the medium should be ascertained before testing. It is also important to determine the effect of the material on the pH of the culture medium, since extremes of pH can be clastogenic (Morita *et al.*, 1989). Buffers can be introduced but in such circumstances a fairly detailed knowledge of the chemistry of the test agent is desirable, since its effectiveness may be strongly pH dependent.

Water or tissue culture medium is the most suitable solvent when testing solid materials but organic solvents such as acetone, dimethyl sulphoxide (DMSO) and dimethylformamide can be used provided that the volume added to the culture medium is not toxic to the cells. Even with water as a solvent, if >10% v/v is added this can be toxic because of tonicity changes and dilution of nutrients. Direct addition of concentrated solutions of test materials may not give an even distribution of the agent; replacement of medium with medium plus agent is preferable and is essential when extended sampling times are used to avoid depletion of nutrients.

A solvent control sample of the same volume as used in the test samples should be included in each experiment unless the solvent is tissue culture medium; this is to check that the prime solvent does not contain any clastogenic impurities. A known clastogen should always be included as a positive control (see also Section 3.3.2.3). If tests are performed with a metabolic activation system, the positive control chemical should be one which requires activation (e.g. cyclophosphamide), to ensure that the system is functioning satisfactorily. If tests are performed without metabolic activation a direct-acting, positive-control chemical should be used (e.g. mitomycin C, methylmethane sulphonate, 4-nitroquinoline-N-oxide). Occasionally it may be possible to select, in addition, a structurally related positive control. Since all tests will involve the use of known clastogens, appropriate safety precautions must be taken (IARC, 1979; MRC, 1981).

3.3 THE PROCEDURE

3.3.1 Basic technique

Details of the procedure are given by Dean & Danford (1984) and Preston *et al.* (1981, 1987). Briefly, cultured cells are treated with the test material in the presence or absence of a metabolic activation system and with appropriate control agents and are harvested at one or more intervals after treatment. Before fixation, cells are exposed to a metaphase-arresting agent such as Colcemid and then to a hypotonic solution to enlarge the cells. They are then fixed in an alcohol/acetic acid solution, dispensed onto microscope slides, stained, randomised, coded and analysed for chromosome abnormalities with high power light microscopy.

3.3.2 Critical factors in the procedure

3.3.2.1 Cell types and kinetics

The cell system used must have been sufficiently validated by showing reproducible sensitivity to known clastogens. Established cell lines, cell strains or primary cell cultures may be employed. The most widely used systems at present are Chinese hamster cell lines and human peripheral blood lymphocytes. The relative merits of these two systems have been discussed by, for example, Kirkland & Garner (1987) and Ishidate & Harnois (1987). We feel it is premature to state a preference.

(a) Chinese hamster cell lines

These are chosen because of their relatively small number of large chromosomes (diploid number, 2n = 22). They may be used at early passage when the chromosome number and karyotype are similar to that *in vivo*, or, more usually, as established cell lines (e.g. Chinese hamster ovary (CHO) cells) in which extensive rearrangement of chromosome material has taken place, the chromosome number may not be constant from cell to cell and the modal number may deviate from 2n = 22. High spontaneous CA frequencies, polyploidy and endoreduplication are sometimes found in these established cell lines but these may be minimised by careful tissue culture techniques such as avoiding over-dense growth of cells by regular subculture, routine checking for mycoplasma infections and avoiding exposure to fluorescent lighting (Parshad *et al.*, 1980). If the spontaneous level of CA (excluding gaps) or polyploidy/endoreduplication exceeds 5 or 10%, respectively, it is most important that the cultures be discarded and new stocks used.

For treatment, cells should be in exponential growth to ensure that there are cells in all stages of the cell cycle (i.e. an asynchronous population). Since the normal cell cycle time is 12–14 hours (Preston *et al.*, 1981) a fixation time of around 20 hours is appropriate (see Section 3.1). If negative or equivocal results are obtained (Section 3.4.4) a repeat study should include the same, and an additional sampling time, approximately 24 hours later. If positive results are obtained, a repeat study need only use the same single sampling time.

(b) Human peripheral blood lymphocytes

Short-term cultures of peripheral blood lymphocytes are stimulated to divide by the addition of a mitogen (e.g. phytohaemagglutinin, PHA) to the culture medium (Evans & O'Riordan, 1975). Mitotic activity begins at about 40 hours after PHA stimulation and reaches a maximum at around 3 days. The chromosome constitution remains diploid during short-term culture.

Treatments should commence at around 48 hours after culture initiation when the cells are actively proliferating and should be sampled first at about 20 hours later, i.e. at 68 hours after culture initiation [the cycle time of lymphocytes, except for the first cycle, averages about 12–14 hours (Morimoto *et al.*, 1983)]. If an additional sampling time is required in a repeat study, because of negative or equivocal results in the first test, it should be at about 24 hours after the first fixation (e.g. 44 hours after the beginning of treatment or 92 hours after culture initiation).

Insufficient data are available to allow us to specify the use of male or female donors, the same or different donors in repeat experiments or separated lymphocytes as opposed to whole blood.

Blood should be obtained from healthy volunteers not receiving medication or knowingly suffering from viral infections. Regular donors should be shown to be hepatitis B antigen negative and staff who regularly handle blood samples should be immunised against hepatitis B. It is recommended that copies of the leaflet *AIDS: Think before you give blood*, issued by the UK National Blood Transfusion Service, be prominently displayed in the workplace of potential donors. Apart from the unavoidable use of hypodermic needles in taking blood samples, sharp instruments should be avoided in the handling of blood cultures to minimise the risk of contracting AIDS. Blood and blood cultures should be handled at containment level 2 as defined by the Advisory Committee on Dangerous Pathogens (1984).

3.3.2.2 Treatment with the test agent

Treatments may be given either continuously up to the time of harvesting the cells or for a few hours duration followed by washing and addition of fresh medium. Short treatments are usually necessary when using an exogenous activation system because such preparations are often cytotoxic when used for extended periods. After such treatments cells must be recultured in fresh medium to allow progression through the cell cycle. Similar reculturing is necessary when treatments have been given in serum-free medium.

The advantage of continuous treatment in complete medium is that it avoids the centrifugation steps which are needed for washing lymphocytes. It will also enable full use to be made of any endogenous metabolic capacity of the test cells. However, the chromosome morphology may be sub-optimal after continuous treatment.

In either case (short or continuous treatment), cells should be sampled at about 1.5 normal cell cycle times from the beginning of treatment in the first test. This sampling may be supplemented with a later time in repeat tests as discussed earlier (Section 3.3.2.1)

3.3.2.3 Controls

A complete test will comprise the following groups:

1. Test material (minimum of three doses) with and without (±) a metabolic activation system.
2. A positive control ± metabolic activation.
3. Solvent only ± metabolic activation.
4. Untreated (negative) control ± metabolic activation.

In practice the negative control (group 4) can usually be omitted when extensive experience has been obtained with a particular cell type, solvent and activation system. Group 3 will be omitted if tissue culture medium is used as a solvent. In repeat tests when two fixation times are used, the positive control will be needed at only one time; the negative or solvent control will be needed at both times.

A chemical can only be regarded as non-clastogenic if it has been tested both with and without metabolic activation. Positive control chemicals should be used at concentrations which induce a relatively low frequency of aberrations so that the sensitivity of the assay for detecting weak clastogens can be established (Preston *et al.*, 1987). Aberration yields in controls should be used to provide a data-base of historical + and − controls.

3.3.2.4 Metabolic activation

Various procedures have been adopted in attempts to convert potential (indirect) mutagens to their active forms *in vitro*. The method most commonly used (Natarajan *et al.*, 1976; Madle & Obe, 1980) in cytogenetic tests involving Chinese hamster cells or human lymphocytes is to add a liver homogenate (S9 plus co-factors), together with the test agent, to the medium in which the cells are growing. Liver S9 is usually obtained from rats which have been Aroclor-induced but other species and inducing agents have been used (e.g. Hubbard *et al*, 1985).

Prolonged treatments with S9-mix may be cytotoxic so the duration of treatment should be as long as possible, consistent with avoiding this problem. Typically 3–6 hour exposures are optimal at between 1 and 10% of the culture medium by volume. Clastogenic effects of Aroclor-induced S9-mix have recently been reported in mouse lymphoma and CHO cells but not in human lymphocytes (Cifone *et al.*, 1987; Kirkland *et al.*, 1989). This may be due to the generation of active oxygen species which are inactivated by blood components. Clastogenesis may be minimised by the use of S9 from animals pre-treated with other enzyme-inducing agents such as phenobarbitone/β-naphthoflavone (Kirkland *et al.*, 1989).

3.3.2.5 Selection of doses

Detectable levels of CA are often found only at doses of clastogens which induce some evidence of cytotoxicity and this may occur over a very narrow dose range. Therefore prior to testing it is necessary to establish the cytotoxicity of the test material in order to select a suitable dose range for the CA assay. Pre-test cytotoxicity assays (Dean & Danford, 1984) will be needed for the test agent alone and the agent plus activation system.

The most common method is to determine the effect of the agent on the mitotic index (MI, percentage of cells in mitosis) at the time at which the cells will be harvested in the CA test. The highest dose chosen for testing should be one which causes a significant reduction in mitotic index but not so great a reduction that insufficient mitotic cells can be found for chromosome analysis. With a control MI of 5–10% it should be possible to score sufficient cells when the MI is reduced by 75% (i.e. the MI is 25% that of the controls).

A minimum of three doses of test agent should be used for the CA assay: the highest dose selected as indicated above, the lowest dose being on the borderline of toxicity, and intermediate doses giving some degree of mitotic inhibition. In practice it is advisable to use more that three doses to ensure that a suitable cytotoxic range is covered and to enable the detection of any dose-response. Cells from only three suitable doses need be analysed. It is extremely important to obtain mitotic index data in the main tests and not just in preliminary studies, to confirm the choice of doses. When mitotic indices are being scored (minimum of 1000 cells per culture) the frequency of polyploid and endoreduplicated cells should also be noted so that the level in control cultures can be monitored (see Section 3.3.2.1 (a)) and their induction by the test agent can be recorded when a late sampling time is used.

If an additional delayed fixation time is required in a repeat test the doses chosen should be those which induce a suitable degree of mitotic inhibition *at the earlier fixation time*; it is unnecessary to use a different range of doses at the second fixation time. Provided that the highest dose used at the first fixation time reduces the MI by approximately 75%, then only this one dose need be scored at the second fixation time. If this highest dose reduces the MI to an unacceptably low level at the second sampling time, giving insufficient metaphases for analysis, the next highest dose should be chosen for scoring, and so on. A negative or solvent control group should also be scored at this second fixation time.

When using cell lines, cytotoxicity can be also assessed by making cell counts at the harvesting time used in the CA test, or by using a colony-forming assay. In the lymphocyte system, total white cell counts can be used in addition to MI. Typically, a dose which induces 50–75% toxicity in these assays will be suitable as the highest test dose but this should be accompanied by a suitable reduction in mitotic index.

A number of highly soluble, non-toxic, non-DNA-reactive substances such as sodium chloride and sucrose have been found to be clastogenic at very high concentrations (>20 mM), probably through changes in the osmolality of the culture medium (Ishidate *et al.*, 1984; Brusick, 1987). To avoid this problem the Japanese Guidelines (see Ishidate, 1988b) recommend an upper concentration limit of 10 mM; we support this recommendation. If tests *are* performed above 10 mM, the osmolality of the culture medium should be measured. If there is a substantial increase (>50 mOsmol/kg) and the chemical nature of the test agent does not suggest DNA reactivity, clastogenesis as a consequence of the high osmolality of the culture medium should be suspected. We do not recommend the testing of chemicals at concentrations above their solubility limits i.e. as suspensions or precipitates.

Duplicate cultures of each test group should be set up except for solvent or negative controls, which should be in quadruplicate to allow the scoring of more cells than in each treated group, as this improves the sensitivity of the assay (Richardson *et al.*, 1989).

3.3.2.6 Scoring methods

Metaphase analysis should only be carried out by a trained and experienced observer who is completely familiar with the karyotype of the cells used for the test and with all the various types of structural changes that can be induced. Prior to scoring, slides should be coded, randomised and then scored 'blind'. Metaphase cells for analysis should be sought under low power magnification and those which are apparently intact and with well spread chromosomes should then be examined under high power (oil immersion). Fixation procedures often result in the breakage of a proportion of metaphases, so in order to have sufficient cells for analysis it is acceptable to analyse cells that have 'lost' one or two chromosomes through processing. For example, human lymphocyte metaphases with 44 or more centromeres can be scored (2n = 46). In CHO cells, if the true range of chromosome numbers is 21–24 (established from preparations in which cell breakage is avoided) then cells with 20 or more centromeres are acceptable. Chromosome numbers should be recorded for each cell; however, polyploid cells should not be scored for chromosome aberrations. This has the added advantage that when a late sampling time is used, allowing cells to pass through more than one cell cycle, indications of aneuploidy induction can be detected. Only hyper-diploid cells should be regarded as such an indicator, i.e. more than 46 chromosomes in human lymphocytes and more than the maximum number in control cultures in CHO cells (>24 chromosomes in the example given above). Cells with less than the expected number of chromosomes may arise as a result of mechanical breakage of metaphases during processing.

If, having selected a cell under low power, it is obvious under high power that the quality of the metaphase precludes accurate CA scoring (e.g. excessively 'fuzzy' or overlapping chromosomes) it should not be included in the scoring. Recording the co-ordinates of cells which are scored is strongly recommended, to allow, if necessary, independent verification by a second observer. Co-ordinates of abnormal cells should always be recorded. It is useful to record representative photographs of chromosomally aberrant cells. Initially a minimum of 200 cells should be scored from each treatment group, 100 from each of two replicates. If ambiguous results are obtained, further 'blind' scoring of these same samples may be undertaken including the additional replicates of negative or solvent controls (Section 3.3.2.5).

3.3.2.7 Extent of testing

Regardless of the result of the first test, it should be repeated. If the first test gives a clear-cut positive result, the repeat study need only utilise the same single sampling time. If the result of the first test is negative or on the borderline of statistical significance the repeat study should use the same, and a later time (see Sections 3.3.2.1 (a) and (b) for specific recommendations) but only the highest dose and a negative or solvent control need be scored at the later time (Section 3.3.2.5).

3.4 DATA PROCESSING AND PRESENTATION

3.4.1 Recording of data

Adoption of the classification and nomenclature of the International System for Human Cytogenetic Nomenclature (ISCN, 1985) as applied to acquired chromosome aberrations is recommended. Some of the most common CA are shown in Table 3.1; the abbreviations used in our earlier Report (Scott *et al.*, 1983), which were based upon Savage (1976), are given together with those used in the ISCN. For a more comprehensive description and discussion of CA types, readers should consult Savage (1983). A suggested score sheet, giving examples of aberrations seen in ten cells is reproduced as Figure 3.1.

From such score sheets the frequencies of various aberration types should be calculated and presented as in Figure 3.2 In calculating the total aberration frequency the practice of counting an exchange as two events and a break as one is not recommended since it makes unwarranted assumptions about the mechanisms involved (see e.g. Revell, 1974). Each aberration should be given equal weight.

3.4.2 Statistical treatment

If a substantial increase in CA frequency is observed with the test agent at at least one dose level, or if there is no indication of any increase at any dose or sampling time, then no statistical analyses of data are necessary. In all other circumstances such analyses should be used.

Comparisons should be made between the frequencies in control cells and at each dose level by using Fisher's Exact test (Richardson *et al.*, 1989).

3.4.3 Presentation of results

This section details the information that should be provided in the test report.

CODE NO OF SLIDE: 27 MICROSCOPE USED: Zeiss 23897 SCORER'S NAME: A. Brown DATE: 11/10/88

Cell No	Chr No	Normal	Gaps	ABERRATIONS[1] Chromosome Deletions[2]	Chromosome Exchanges[3] (cse)	Chromatid Deletions[2]	Chromatid Exchanges[3] (cte)	Others	Vernier	Diagram/Comments
1	46	N							94/49.3	
2	46		csg						94.4/53.1	
3	45			ace	(dic+ace)				96.1/53.7	One ace allocated to the dic
4	46		ctg			ctb, Nup			96.3/46.1	Acentric fragment of Nup remains aligned
5	46			min	(r+ace)				97/44.3	
6	46							pvz	98.3/50.5	
7	44					min	r		103.4/51.9	r
8	46		2ctg				qr, tr(2cen)		101.3/41.4	qr, Asymmetrical; c tr(2cen)
9	46				(tri+2ace)				103.8/37.4	= (tri+2ace)
10	46		ctg			ctb			107/58.5	
Σ 10		1	5	2	4	4	3	1		

Fig. 3.1. Typical score sheet (10 cells).

Notes:

1. ISCN (1985) abbreviations used except for Nup because no abbreviation is given for isochromatid breaks.
2. It is often difficult to distinguish between the various intra-arm events [e.g.acentric fragments (ace), minutes (min) and isochromatid breaks] so for convenience these are listed as deletions although they may be derived from exchange events. Inter-arm and inter-chromosome events are listed under exchanges.
3. Each exchange should be counted as just one aberration (see Section 3.4.1).

ace = acentric fragment
csg = chromosome gap
ctg = chromatid gap
ctb = chromatid break
dic = dicentric
(dic + ace) = dicentric with associated acentric fragment. Bracketted to indicate that it is a single aberration
min = minute. There are chromatid minutes (ct min) and chromosome minutes (cs min)
N = normal cell, no aberrations
Nup = isochromatid break, incomplete proximally
pvz = pulverised chromosomes with very many gaps and breaks, not associated with exchanges. Too many aberrations to count. In totalling the aberrations these cells are normally ignored but are included under '% cells with aberrations' (see Fig. 3.2)
qr = quadriradial, a chromatid exchange (cte)
r = ring chromosome (with centromere), chromosome (csr) or chromatid (ctr) type
(r + ace) = ring with associated acentric fragment. Bracketted to indicate that it is a single aberration
tr = triradial
tr(2cen) = triradial with 2 centromeres
(tri + 2ace) = tricentric with 2 associated acentric fragments. This should be counted as 2 aberrations, i.e. equivalent to 2 dic + ace

Treatment	Replicate No	No cells scored	ABERRATIONS PER 100 CELLS								% Cells with abs (+gaps)	% Cells with abs (-gaps)	MI
			Gaps	Chromosome		Chromatid		Others	Total abs (+gaps)	Total abs (-gaps)			
				Deletions	Exchanges	Deletions	Exchanges						
Neg. control	1												
Neg. control	2												
	Total												
Pos. control	1												
Pos. control	2												
	Total												
Dose 1	1												
Dose 1	2												
	Total												
Dose 2	1												
Dose 2	2												
	Total												
Etc.													

Fig. 3.2. Data summary sheet.

3.4.3.1 Test material

See Chapter 1 of these Guidelines.

3.4.3.2 Test cells

Name (e.g. Chinese hamster ovary, giving source of the cells; human peripheral blood lymphocytes, giving age and sex of donor), culture medium (giving proportions of components), amount of blood per culture and whether whole blood or separated lymphocytes (in latter case give cell numbers per ml of culture medium), type of culture vessel, temperature of incubation and gas conditions (e.g. 5% CO_2).

3.4.3.3 Method of treatment

For cell lines, the time of subculturing prior to treatment and the number of cells plated should be given and, for blood cultures, the time between mitogen stimulation and treatment (e.g. 48 hours). Also required is the volume of medium in which cells are treated, volume of vehicle + test agent which is added, how it is added (direct to cell cultures or pre-mixed in medium/saline, etc.), doses used, duration of treatment (e.g. continuous or 3 hours, any centrifugation and/or washing after treatment?) and details of S9-mix (preparation, volume and concentration added, whether added at same time as test material, duration of treatment). The effect of the test agent on the pH of the treatment medium should be noted and the effect on osmolality if high concentrations are used (Section 3.3.2.5).

3.4.3.4 Harvesting cells

Time after treatment, duration and concentration of Colcemid treatment, type of hypotonic solution, fixation and stain.

3.4.3.5 Cytotoxicity assay

Method used (e.g. mitotic index, cell counts, colony-forming ability), doses used and results obtained.

3.4.3.6 Data presentation

For each replicate of each of the treatment groups specified in Section 3.3.2.3 (i.e. controls and each dose of test agent) the number of cells scored, percentage with aberrations, the calculated frequency per 100 cells of gaps, chromosome and chromatid breaks and exchanges, and the MI and other indicators of toxicity (if measured) should be tabulated. A suggested layout is given in Figure 3.2.

3.4.3.7 Interpretation

A clear statement of whether or not the test agent is considered to be clastogenic should be given (with reasons). If statistical analyses have been used, details should be given.

The following Section (3.4.4) deals with the designation of a positive or negative response.

3.4.4 Interpretation of data as a positive or negative response

It may be possible to designate the result of a test as being + or − without resort to statistical analyses (see Section 3.4.2). However, a situation that arises quite frequently and causes difficulties in interpretation, is the absence of a clear positive dose-response relationship at a particular sampling time. This should not really be a cause for concern since it is entirely to be expected if a single, common, sampling time is used for all doses of test agent. It arises because CA yields usually vary markedly with post-treatment sampling time of an asynchronous population and because increasing doses of clastogens usually induce increasing degrees of mitotic delay. This point is illustrated in Figure 3.3, from which it can be seen that positive, negative or intermediate dose-response curves can be obtained in a situation in which the aberration yield in the whole population shows a simple positive dose-response. The use of additional fixation times will often clarify the relationship between dose and aberration yield.

With these considerations in mind it is unwise to insist on a positive dose-response at a particular sampling time before designating a result as being positive. Provided that at least one dose at one fixation time gives a yield which is so substantially greater than the historical negative control range that statistical analysis is unnecessary, the result should be accepted as positive (Richardson *et al.*, 1989). An ambiguous situation may arise when no individual dose gives a yield which is significantly above controls (P not <0.05) but the combined data for treated cells against controls are significant. Such situations are often resolved when an additional sampling time is used.

Although gaps are traditionally excluded from quantitation of CA yields, they should not be ignored in equivocal situations. One good reason for this is that, even when there is no misalignment of the affected chromatid, some gaps have been shown to be real discontinuities in DNA (Heddle & Bodycote, 1970; Satya-Prakash *et al.*, 1981). Where

CA yields are on the borderline of statistical significance above controls and the inclusion of gaps makes the yields clearly significant, further investigations are required. Since some lesions scored as gaps are, in reality, true breaks, then an agent inducing significant numbers of gaps but not of other aberration types must be regarded as clastogenic unless it can be shown that the types of gaps induced are *not* discontinuities. The simplest way to show this would be to demonstrate no significant increase in micronuclei or anaphase fragments (Countryman & Heddle, 1976; Nichols *et al.*, 1984).

Chromosomal exchanges are comparatively rare spontaneous events and have important genetic consequences (see Section 3.1). In borderline situations, therefore, greater significance should be attached to the observation of exchanges in treated cells than to a small numerical increase in gaps and breaks.

Fig. 3.3. The influence of mitotic delay on the shape of dose-response curves (lower part of figure) when a single sampling time is used. The type of dose-response curve obtained will depend on the relationship between aberration yield and time, for each of the doses used (upper part of figure). The three curves shown are for times when the maximum yield is induced by the lowest, middle and highest dose, respectively.

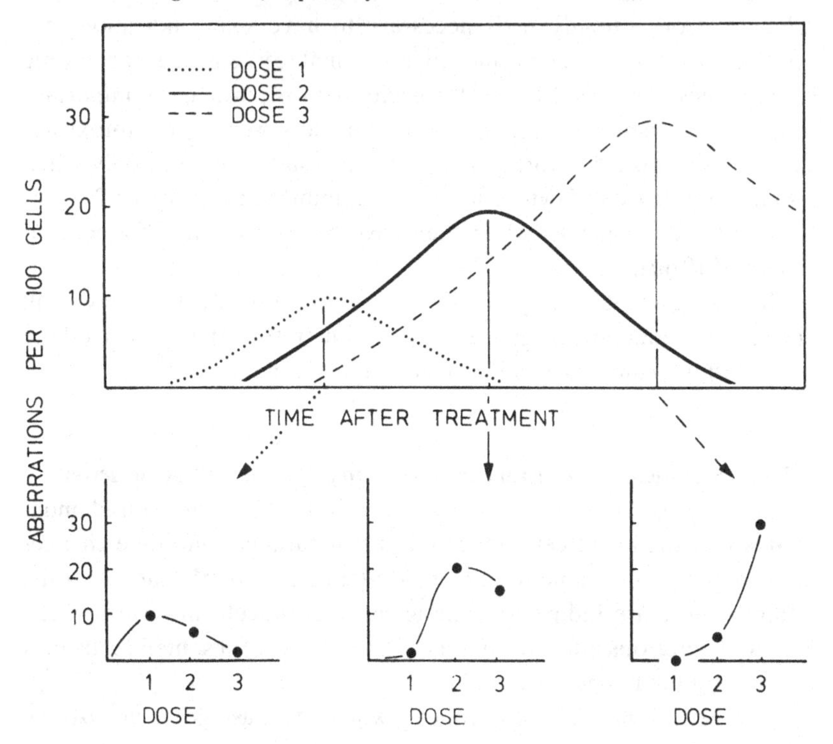

Consideration cannot be given, in this chapter, to all conceivable types of test result. The ultimate designation must rely upon experience and sound scientific judgement.

3.5 DISCUSSION

3.5.1 Critical factors in the assay

The most critical factor determining the validity of a CA assay is the competence of the scorer. Metaphase analysis of structural chromosome aberrations requires a considerable degree of skill, necessitating good training and extensive experience. Coupled to this is the need for high quality chromosome preparations allowing unambiguous scoring of aberrations. No attempt should be made to analyse poor quality metaphases.

The choice of doses in relation to cytotoxicity is another critical factor. Since many chemicals are clastogenic only at cytotoxic doses, the extent of cytotoxicity must be determined. Even though toxicity assays are performed prior to the actual clastogenicity test, to determine the dose range, the same degree of toxicity may not actually be achieved in the final test (repeatability of cytotoxicity levels is notoriously difficult to achieve). Consequently, it is necessary to have some indication that cytotoxic levels have been achieved in the final test, e.g. reduced mitotic index values. The need for such toxicity data is particularly important in claiming a negative response since, in the absence of cytotoxicity, such a claim may be worthless. Negative results are valid only when a significant degree of cytoxicity has been induced (Section 3.3.2.5) or, for non-toxic chemicals, when they have been tested up to a concentration of 10 mM.

The use of a single sampling time will not necessarily detect the peak aberration frequency. Negative or borderline results therefore need the repeat study to include an additional sampling time.

3.5.2 Significance of results in relation to intrinsic mutagenic activity

A positive result in a CA assay in cultured somatic cells demonstrates that the agent can induce gross structural chromosome changes *in vitro* and *may* be able to induce non-lethal structural changes. If the latter changes are induced by the agent in germ cells *in vivo* they can lead to deleterious effects in embryos and progeny; in somatic cells they may contribute to neoplastic changes.

However, about 50% of chemicals which are clastogenic *in vitro* fail to induce CA when tested *in vivo* (Thompson, 1986; Ishidate *et al.*, 1988). There are many possible reasons for this (Waters *et al.*, 1988), including the fact that CA can be induced by extreme culture conditions [e.g. increased osmolality at high concentrations, or low pH, etc. (Brusick, 1987)].

Chemicals which are clastogenic *in vitro* at low concentrations are more likely to be clastogenic *in vivo* than those whose clastogenicity is detected only at high concentrations. However, there are exceptions and some *in vivo* positives require high *in vitro* concentrations for detection; this is probably largely a reflection of the inadequacy of metabolic activation systems *in vitro*. Nevertheless, if an upper concentration limit of 10 mM is adopted, to avoid the artefactual clastogenesis which can occur at higher concentrations (Section 3.3.2.5), most, if not all chemicals with a potential for *in vivo* clastogenesis will be detected (Scott *et al.*, 1990).

Negative results in well conducted *in vitro* tests are a strong indication of lack of potential for *in vivo* clastogenesis since almost all *in vivo* clastogens have given positive results *in vitro* when adequately tested (Thompson, 1986; Ishidate *et al.*, 1988).

3.6 CONCLUSIONS

1. This chapter considers the use of metaphase analysis of structural chromosome aberrations in cultured mammalian somatic cells as a screening method for the detection of clastogens and carcinogens. Many of the gross structural changes observed will be cell lethal but are taken as an indication of non-lethal structural changes which have important consequences in both germ and somatic cells.
2. The most widely used cell systems are Chinese hamster cells and short-term cultures of human peripheral blood lymphocytes. These should be treated when actively proliferating and sampled at approximately 1.5 normal cell cycle times from the beginning of treatment.
3. Regardless of the result of the first test it should be repeated. If the result of the first test is negative or on the borderline of statistical significance, the repeat study should utilise the same and a later sampling time.
4. A minimum of three doses of test agent should be used together with appropriate positive and negative controls. In a repeat test with two sampling times only the highest dose and a negative or solvent control need by scored at the later time.

5. Pre-test cytotoxicity assays are required to establish the dose range. Final tests must include cytotoxic doses up to 10 mM. Cytotoxicity should be measured in the final tests.
6. The International System for Human Cytogenetic Nomenclature of chromosome aberrations is strongly recommended.
7. Fisher's Exact tests should be used where statistical analysis is required. At a particular sampling time, simple positive dose/response curves may not always be obtained.
8. The Test Report should give detailed information on methods used, and a breakdown of aberrations into gaps, chromosome and chromatid breaks and exchanges.

3.7 REFERENCES

Advisory Committee on Dangerous Pathogens (1984). *Categorisation of Pathogens according to Hazard and Categories of Containment*. HMSO, London.

Bender, M.A., Griggs, H.G. & Bedford, J.S. (1974). Mechanisms of chromosomal aberration production. III. Chemicals and ionising radiation. *Mutation Research*, **23**, 197–212.

Brusick, D. (ed.) (1987). Genotoxicity produced in cultured mammalian cell assays by treatment conditions. *Mutation Research*, **189**, 1–80.

Chandley, A.C. (1981). The origin of chromosomal aberrations in man and their potential for survival and reproduction in the adult human population. *Annals of Genetics*, **24**, 5–11.

Cifone, M.A., Myhr, B., Eiche, A. & Bolcsfoldi, G. (1987). Effect of pH shifts on the mutant frequency at the thymidine kinase locus in mouse lymphoma L5178Y TK+/– cells. *Mutation Research*, **189**, 39–46.

Countryman, P.I. & Heddle, J.A. (1976). The production of micronuclei from chromosome aberrations in irradiated cultures of human lymphocytes. *Mutation Research*, **41**, 321–32.

Dean, B.J. & Danford, N. (1984). Assays for the detection of chemically-induced chromosome damage in cultured mammalian cells. In *Mutagenicity Testing: A Practical Approach*, ed. S. Venitt and J.M. Parry. IRL Press, Oxford, pp. 187–232.

European Communities (1984). *Official Journal of the European Community, L251, Vol. 27*, pp. 132.

Evans, H.J. & O'Riordan, M.L. (1975). Human peripheral blood lymphocytes for the analysis of chromosome aberrations in mutagen tests. *Mutation Research*, **31**, 135–48.

Heddle, J.A. & Bodycote, D.J. (1970). On the formation of chromosomal aberrations. *Mutation Research*, **9**, 117–26.

Hubbard, S.A., Brooks, T.M., Gonzalez, L.P. & Bridges, J.W. (1985). Preparation and characterisation of S9 fractions. In *Comparative Genetic Toxicology*, ed. J.M. Parry and C.F. Arlett. The Macmillan Press Ltd, Houndmills, Hants, pp. 413–38.

IARC (1979). Handling chemical carcinogens in the laboratory; problems of safety. *Scientific Publication No. 33*, International Agency for Research on Cancer, Lyon.

ISCN (1985). *An International System for Human Cytogenetic Nomenclature*, ed. D. G. Harnden *et al*. S. Karger, Switzerland.

Ishidate, M., Jr (1988a). *Data Book of Chromosomal Aberration Tests* in vitro. Elsevier, Amsterdam.

Ishidate, M. Jr (1988b). A proposed battery of tests for the initial evaluation of the mutagenic potential of medicinal and industrial chemicals. *Mutation Research*, **205**, 397–407.

Ishidate, M. Jr & Harnois, M.C. (1987). The clastogenicity of chemicals in mammalian cells. Letter to the Editor. *Mutagenesis*, **2**, 240–3.

Ishidate, M. Jr, Harnois, M.C. & Sofuni, T. (1988). A comparative analysis of data on the clastogenicity of 951 chemical substances tested in mammalian cell cultures. *Mutation Research*, **195**, 151–213.

Ishidate, M. Jr, Sofuni, T., Yoshikawa, K., Hayashi, M., Nohmi, T., Sawada, M. & Matsooka, A. (1984). Primary mutagenicity screening of food additives currently used in Japan. *Food and Chemical Toxicology*, **22**, 623–36.

Kirkland, D.J. & Garner, R.C. (1987). Testing for genotoxicity – chromosomal aberrations *in vitro* – CHO cells or human lymphocytes? *Mutation Research*, **189**, 186–7.

Kirkland, D.J., Marshall, R.R., McEnaney, S., Bidgood, J., Rutter, A., & Mullineux, S. (1989). Aroclor-1254 induced rat liver S-9 causes chromosome aberrations in CHO cells but not in human lymphocytes: A role for active oxygen? *Mutation Research*, **214**, 115–22.

Madle, S. & Obe, G. (1980). Methods for analysis of the mutagenicity of indirect mutagens/carcinogens in eukaryotic cells. *Human Genetics*, **56**, 7–20.

Minden, M.D. (1987).Oncogenes. In *The Basic Science of Oncology*, ed. I.F. Tannock and R.P. Hill. Pergamon Press, New York, pp. 72–88

Morimoto, K., Sato, M. & Koizumi, A. (1983). Proliferative kinetics of human lymphocytes in culture measured by autoradiography and sister chromatid differential staining. *Experimental Cell Research*, **145**, 249–56.

Morita, T., Watanabe, Y., Takeda, K. & Ukumura, K. (1989). Effects of pH in the *in vitro* chromosomal aberration test. *Mutation Research*, **225**, 55–60.

MRC (1981). *Guidelines for work with Chemical Carcinogens in Medical Research Council Establishments*. Medical Research Council, London.

Natarajan, A.T., Tates, A.D., van Buul, P.P.W., Meijers, M. & de Vogel, N. (1976). Cytogenetic effects of mutagens/carcinogens after activation in a microsomal system *in vitro*. *Mutation Research*, **37**, 83–90.

Nichols, W.W., Miller, R.C. & Bradt, C.I. (1984). *In vitro* anaphase and metaphase preparations in mutation testing. In *Handbook of Mutagenicity Test Procedures*, ed. B.J. Kilbey, M. Legator, W. Nichols and C. Ramel. Elsevier Science Publishers BV, Amsterdam, pp. 429–39.

Parshad, R., Taylor, W.G., Sanford, K.K., Camalier, R.F., Gantt, R. & Tarone, R.E. (1980). Fluorescent-light-induced chromosome damage in human IMR-90 fibroblasts. Role of hydrogen peroxide and related free radicals. *Mutation Research*, **73**, 115–24.

Perry, P., Henderson, L. & Kirkland, D. (1984). Sister chromatid exchange in cultured cells. In *UKEMS Sub-committee on Guidelines for Mutagenicity Testing. Report. Part II. Supplementary Tests*, ed. B.J. Dean. United Kingdom Environmental Mutagen Society, Swansea, pp. 89–121.

Phillips, R.A. (1987). The genetic basis of cancer. In *The Basic Science of Oncology*, ed. T.F. Tannock and R.P. Hill. Pergamon Press, New York, pp. 24–51.

Preston, R.J., Au, W., Bender, M.A., Brewen, J.G., Carrano, A.C., Heddle,

J.A., McFee, A.F., Wolff, S. & Wassom, J.S. (1981). Mammalian *in vivo* and *in vitro* cytogenetic assays. *Mutation Research*, **87**, 143–88.

Preston, R.J., San Sebastian, J.R. & McFee, A.F. (1987). The *in vitro* human lymphocyte assay for assessing the clastogenicity of chemical agents. *Mutation Research*, **189**, 175–83.

Revell, S.H. (1974). The breakage-and-reunion theory and the exchange theory for chromosome aberrations induced by ionising radiations: A short history. In *Advances in Radiation Biology, Vol. 4*, ed. J.T. Lett, H. Adler and M. Zelle. Academic Press, New York, pp. 367–415.

Richardson, C., Williams, D.A., Allen, J.A., Amphlett, G., Chanter, D.O. & Phillips, B. (1989). Analysis of data from *in vitro* cytogenetic assays. In *UKEMS Sub-committee on Guidelines for Mutagenicity Testing. Report. Part III. Statistical Evaluation of Mutagenicity Test Data*, ed. D. J. Kirkland. Cambridge University Press, Cambridge, pp. 141–54.

Sankaranarayanan, K. (1982). *Genetic effects of ionising radiation in multicellular eukaryocytes and the assessment of genetic radiation hazards in Man*. Chapter IV. Chromosome Aberrations. Elsevier Biomedical Press, Amsterdam, pp. 151–257.

Satya-Prakash, K.L., Hsu, T.C. & Pathak, S. (1981). Chromosome lesions and chromosome core morphology. *Cytogenetics and Cell Genetics*, **30**, 248–52.

Savage, J.R.K. (1976). Annotation: Classification and relationships of induced chromosome structural change. *Journal of Medical Genetics*, **13**, 103–22.

Savage, J.R.K. (1983). Some practical notes on chromosomal aberrations. *Clinical Cytogenetics Bulletin*, **1**, 64–76 (Obtainable from the author at MRC Radiobiology Unit, Chilton, Didcot, Oxon. OX11 0RQ).

Scott, D., Danford, N., Dean, B., Kirkland, D. & Richardson, C. (1983). *In vitro* chromosome aberration assays. In *UKEMS Sub-committee on Guidelines for Mutagenicity Testing. Report. Part I. Basic Test Battery*, ed. B.J. Dean. United Kingdom Environmental Mutagen Society, Swansea, pp. 41–64.

Scott, D., Galloway, S. M., Marshall, R. R., Ishidate, M. Jr, Brusick, D., Ashby, J. & Myhr, B. C. (1990). Genotoxicity under extreme culture conditions: A report from ICPEMC Task Group 9. *Mutation Research* (in press).

Thompson, E.D. (1986). Comparison of *in vivo* and *in vitro* cytogenetic assay results. *Environmental Mutagenesis*, **8**, 753–67.

Thust, R., Mendel, J., Schwarz, H. & Warzok, R. (1980). Nitrosated urea pesticide metabolities and other nitrosamides. Activity in clastogenicity and SCE assays, and aberration kinetics in Chinese hamster V79-E cells. *Mutation Research*, **79**, 239–48.

Waters, M.D., Bergman, H.B. & Nesnow, S. (1988). The genetic toxicology of Gene-Tox non-carcinogens. *Mutation Research*, **205**, 139–82.

4

Gene mutation assays in cultured mammalian cells

J. COLE
M. FOX
R.C. GARNER
D.B. McGREGOR
J. THACKER

4.1 INTRODUCTION

4.1.1 Genetic basis of assay

The use of cultured mammalian cells, including human cells, for mutation detection gives a measure of the intrinsic response of the mammalian genome to mutagens, while offering rapidity of assay and ease of treatment when compared to the use of whole mammals. Over the last 5 years our knowledge of mammalian cell mutation has improved dramatically, especially in terms of the underlying molecular changes. Additionally there has been a consolidation of data with other mutation assays and the introduction of new systems using human cells.

While several mutation systems are available for mammalian cells, few have been defined adequately for quantitative studies. The majority of such systems work by placing a mutagen-treated cell population under selective pressure (e.g. exposure to a toxic drug) so that only mutant cells are able to survive. Mutation of genes in a hemizygous (single copy) or heterozygous (two copies but only one active) state is usually studied, to enable mutations to be measured at an adequate frequency. Thus, favoured systems exist for mutation of X-chromosome genes (where only one X is present in male cells, and only one active X in female cells) or for autosomal genes which have been identified to be in the hemi- or heterozygous state in certain cell lines.

Commonly used systems work by selecting for the loss of function of a gene product (enzyme) which is inessential for the survival of cells in culture. For example, in mammalian cells certain enzymes 'salvage' nucleic acid breakdown products (purine and pyrimidine bases) for re-use in metabolism, although the cell has the capability to synthesise bases *de novo*. Additionally, if toxic base analogues are put into the growth medium the corresponding salvage enzyme will incorporate the analogue into the cells. Thus the cells will die in the presence of a sufficient concentration of the analogue unless the salvage enzyme is rendered inactive (by mutation, for example). Examples of such systems are: resistance to 6-thioguanine (6TG) resulting from lack of hypoxanthine phosphoribosyl transferase (HPRT) enzyme activity (e.g. Bradley *et al.*, 1981; Li *et al.*, 1987), resistance to trifluorothymidine (TFT) from lack of thymidine kinase (TK) enzyme activity (e.g. Liber & Thilly, 1982; Clive *et al.*, 1987), resistance to 8-azaadenine or to 2,6-diaminopurine from lack of adenine phosphoribosyl transferase (APRT) activity (e.g. Thompson *et al.*, 1980; Steglich & De Mars, 1982; Paeratakul & Taylor, 1986; McKenna & Ward, 1987; Ward & McKenna, 1987). Since complete loss of function of these enzymes is not deleterious to the cultured cell, in theory all types of mutant from those with simple base substitutions to those with large deletions or rearrangements should be detectable. However, molecular analyses of the genetic changes in analogue-resistant cells have recently defined several important features of these systems.

1. The selective conditions used *are* successful in allowing the detection of genetically altered cells of various types, i.e. all changes examined in detail appear to be classifiable as mutations in the accepted sense. When defining variant clones selected *in vitro* as 'mutant', it should be remembered that in some systems, stable loss of gene function has been shown to be due to hypermethylation of the genetic region rather than alteration of the sequence or structure of the gene (e.g. Gebara *et al.*, 1987; Dubrovic *et al.*, 1988). However, in the systems commonly used for mutation tests, hypermethylation has not as yet been found to be responsible for the mutant phenotype (e.g. Vrieling *et al.*, 1988; Moore *et al.*, 1988).

However,

2. each system appears to detect a different range of genetic changes (although these ranges overlap);
3. some of the systems have been shown to discriminate between the actions of different mutagens (i.e. where a range of mutant types is detected, some mutagens induce almost exclusively point mutations while others induce predominantly larger changes);
4. the types of mutation detected are comparable to those found to activate human proto-oncogenes in tumour cells (Barbacid, 1986); additionally, there is some evidence that the spontaneous frequency of different types of mutations detected by these

somatic cell systems is similar to that at the same locus in mammalian germinal tissue (e.g. for HPRT-deficiency, compare Yang *et al.*, 1984; Stankowski *et al.*, 1986; Bradley *et al.*, 1987; Thacker & Ganesh, 1989).

On the question of the ability of the different systems to detect a range of mutation types (points (2) and (3) above), some simple rules are emerging. Where a system uses a cell line which is hemizygous for the test gene (only one copy of the gene and perhaps of its flanking regions in the genome) there is the likelihood that large genetic changes will not be tolerated. That is, if the genetic changes extend to regions which are essential to cell survival, there may be no second copy of those regions available for recombination and the cell will die. The *hprt* gene is X-chromosome-linked (see above) and, therefore, is expected to be restricted in ability to show very large genetic changes (Hsie, 1987; Moore *et al.*, 1987a). However, *hprt* is also a large gene (Melton *et al.*, 1984) and changes of 30–40 kilobases, as well as point mutations, have been identified (Brown & Thacker, 1984; Thacker, 1985; Vrieling *et al.*, 1985; Brown *et al.*, 1986; Fuscoe *et al.*, 1986). The *aprt* gene is much smaller and in hemizygous cells selection for APRT-deficiency detects many fewer large deletions than selection for HPRT deficiency, probably because essential genes are located downstream of *aprt* (e.g. Breimer *et al.*, 1986). Similarly it has been found that cells which are hemizygous for the *tk* gene show a much reduced mutation frequency compared to cells which are heterozygous for the same gene (Evans *et al.*, 1986; Moore *et al.*, 1987b). Two types of colonies are seen when mutations at the *tk* locus are selected in the heterozygous mouse lymphoma cell line L5178Y $TK^{+/-}$: large colonies, which grow at the normal rate, and small, slow growing colonies. Such large and small mutant colonies may be quantitated using an automatic colony counter, and show a bimodal distribution (Moore *et al.*, 1985a, b). Research in recent years has indicated that large-colony TK mutations represent events within the gene (base-pair substitutions or deletions) that affect the expression of the *tk* locus, whereas small-colony mutants carry large genetic changes involving chromosome 11b, the chromosome which carries the active *tk* gene, which are frequently visible as chromosome aberrations using banding techniques (Hozier *et al.*, 1981, 1985; Moore *et al.*, 1985a, b; Applegate & Hozier, 1987; DeMarini *et al.*, 1989). A similar situation may exist in TK6 human lymphoblastoid cells (Yandell *et al.*, 1986; Liber *et al.*, 1989). Thus, in these heterozygous cell lines, both gene mutations within the *tk* gene, and chromosomal events involving the gene, may be detected by selecting TK mutants in medium containing trifluorothymidine (TFT). However, *tk* is a moderate-sized gene (11–13 kilobases) so that, even in the hemizygous state, its use should allow relatively large changes to be detected (Liber *et al.*, 1986). Thus the copy number, size and genomic position of the gene are largely responsible for the detection limits of systems based on that gene.

Additionally, the detection capability of a system may vary widely if the test gene(s) affects the chances of cell survival in a manner different from that of the salvage enzyme genes listed above. This aspect may be illustrated by selection for ouabain-resistance on the one hand and by the use of cells carrying an accessory chromosome on the other. Ouabain kills cells by binding to the essential Na^{+} K^{+}-dependent ATPase enzyme, causing an imbalance in the ion flow. Mutants which fail to bind ouabain while retaining the ionic balance can be selected at relatively low frequency (Baker *et al.*, 1974). Thus it is expected that large mutagenic changes would not be tolerated, since these would destroy the enzyme's essential function, and this view is supported by molecular analyses of ouabain-resistant mutants (Muriel *et al.*, 1987). In contrast, a system based on the inclusion of a single human chromosome in a hamster cell yields very high mutant frequencies for genes on that chromosome: the mutants tolerate large genetic losses, including the whole human chromosome, since it is inessential for cell survival (Waldren *et al.*, 1979).

The majority of systems require colony formation to detect mutants; this has the benefit of allowing further analysis of the selected clones to verify their mutational origins and give confidence in the system (e.g. in human lymphocytes mutation has progressed from single cell to clonal analyses; Albertini *et al.*, 1982, 1985; Morley *et al.*, 1985; Henderson *et al.*, 1986; Cole *et al.*, 1988). Some progress has been made with potentially rapid single-cell systems using antibody markers and flow-cytometric sorting (Langlois *et al.*, 1987). Additionally, the last 5 years have seen the development of a number of shuttle-vector systems in which defined genes are introduced into mammalian cells for mutation induction and/or processing but are then rescued for analysis in bacteria (DuBridge & Calos, 1987). The facility to replicate the target genes in both mammalian cells and bacteria gives a more rapid and precise means of identifying mutations, but it is not yet certain that shuttle systems yield the same spectrum of mutations as that found for 'native' mammalian genes.

4.1.2 Limitations

The generation of satisfactory mutation data from cultured rodent or human cells depends greatly on good quality control. Rodent and human tumour cell lines *in vitro* are genetically unstable and may show drift in sensitivity to mutagens (Fox, 1985a); to ensure reproducibility it is therefore important that large stocks of a well characterised clone be established before starting a series of experiments. Cells should be karyotyped upon receipt, and periodically checked to ensure stability. Cells should also be routinely checked for the absence of mycoplasma and for their spontaneous mutant frequency. Different serum batches vary in their content of nucleosides, e.g. thymidine and hypoxanthine, so it is important that they are pre-screened to ensure stringency of selection as well as for ability to support good growth rates and plating efficiencies. Well established proven protocols are also of prime importance if data are to be comparable between laboratories.

An additional limitation of *in vitro* assays is the restricted metabolic capacity of many cultured cell lines. The third UKEMS comparative genotoxicity trial was instigated specifically to investigate definitive protocols for gene mutation assays using cultured mammalian cells. The provision of recommendations for optimal levels of liver S9 fractions was included among the aims of this study.

A number of laboratories collaborated in the UKEMS study using previously agreed protocols and a detailed examination of co-factor toxicity and S9/co-factor optimisation for Chinese hamster, mouse lymphoma and human lymphoblastoid cells was undertaken. The results have been published in *Mutagenesis* (Supplement to Volume 5, 1990; see also Section 4.3.2.5). It should be noted that S9 itself may be mutagenic when used for prolonged periods (18–24 hours) or at high concentrations (50 μl S9/ml) as may be medium in which S9 has been pre-incubated (Myhr & Mayo, 1987; McGregor *et al.*, 1988).

4.2 THE TEST MATERIAL

4.2.1 Handling procedures for the test material

It is essential that solutions of the test material be freshly prepared; any unused solution should be disposed of safely. If the substance is a volatile solvent or a gas, the boiling point and vapour pressure should be given and the assay procedures modified (see Section 4.2.2.2). If a known impurity is present in the test material, it should be assayed for mutagenicity at doses equivalent to those which would be present in the chosen doses of the major constituent, so that the contribution of the minor component can be assessed. If a mixture is to be tested, this should be stated and details of the preparation provided if possible.

4.2.2 Solvent selection

4.2.2.1 Testing of solids or non-volatile liquids

Ideally, the test material should be dissolved just prior to testing in the medium to be used during the test period. Most materials routinely tested are not soluble in tissue culture medium. A commonly used solvent is anhydrous dimethyl sulphoxide (final concentration in the assay should not exceed 1% v/v). Methanol and ethanol should be avoided with S9-mix because both formaldehyde and acetaldehyde are genetically active. Possible interactions between the solvent and the test substance should also be borne in mind. For a discussion of the use of organic solvents with V79 cells, see Abbondandolo *et al.* (1980).

4.2.2.2 Testing of volatile liquids and gases

These represent considerable problems in terms of mutagenicity testing. Volatile liquids should be dispensed in a fume cupboard and tested in enclosed containers to prevent escape of the test material. Douglas McGregor has developed and evaluated a safe method for testing volatile liquids and gases in mouse lymphoma and other suspension cultures. Details may be obtained from Dr W. J. Caspary, NIEHS, PO Box 12233, Research Triangle Park, NC, USA.

4.2.3 Control chemicals

4.2.3.1 Negative controls

These should consist of the solvent used to dissolve the test chemical at the same final concentration of the solvent. The investigator should be able to show that the solvent used at the given concentration has no toxic or mutagenic effect.

4.2.3.2 Positive controls

Two separate positive controls are necessary, one of which should require metabolic activation. Positive control chemicals of known standard purity are recommended so that for quality control purposes a well defined response can be obtained for assessment of the performance of the assay. Ethylmethane sulphonate (EMS) (note that EMS is hygroscopic and should be stored in sealed ampoules) or 4-nitroquinoline-N-oxide (4-NQO) are suggested as mutagens not requiring exogenous metabolism. For the induction of TK mutants in L5178Y TK$^{+/-}$ cells, hycanthone methane sulphonate or methylmethane sulphonate are recommended as both are potent inducers of small-colony mutants. 3-methylcholanthrene, cyclophosphamide or benzo(a)pyrene [B(a)P] are recommended as indirect-acting compounds. Propylene oxide and vinylidene chloride are recommended as volatile direct- and indirect-acting positive controls. Where the structure of the chemical is known and is related to a known carcinogen then a positive control of the appropriate chemical class should be used. However, we note that it is important to avoid making assumptions about the similarity of the active groups in the positive control and the test chemical or their metabolites, which may not be justifiable.

4.3 THE PROCEDURE

4.3.1 Outline of the basic technique

4.3.1.1 Summary of technique

An outline of the basic techniques, and detailed protocols for individual cell lines, are available in papers quoted in the text, and in the extensive literature cited in reviews (see Section 4.3.1.3).

Briefly, a large population of exponentially growing cells in the presence or absence of a metabolic activation system is treated with the test substance for a limited period of time. After removal of the test substance the cells are washed and a sample diluted to determine cloning efficiency (CE) immediately after treatment. We strongly recommend that this should be done on at least one occasion if accurate estimates of the toxic effect of the chemical are to be made. Estimates based only on daily population counts during the expression time and CE at the time of selection (Clive *et al.*, 1979) may not provide sufficiently accurate data. After treatment, cells are cultured for sufficient time to allow newly induced mutations to be fixed and expressed (the expression time). The term expression time as used in the literature is ambiguous. It can mean either the time of maximum induced mutant frequency (IMF) or the time between treatment and plating into selective media. In many assay systems an approximate plateau level is demonstrable after maximum IMF is reached. When a single expression time is used it should represent maximum induced mutant frequency. During this time the growth rate may be monitored and the cells subcultured if necessary. At the end of the expression time, which must be well defined for each cell line and selective system, samples are cloned in selective and non-selective medium to determine the mutation frequency per survivor.

4.3.1.2 Cell types

Many different cell types, for example S49, 3T3, C3H10T½, and L5178Y mouse cells, ARL6 and other rat liver lines, BHK and SHE Syrian hamster and V79 and CHO Chinese hamster cells, normal and repair-deficient skin-derived human fibroblasts, their SV-40 transformed derivatives and human EBV transformed lymphoblastoid cells, e.g. TK6 (Jensen & Thilly, 1986) and a wide variety of selective systems (Gupta, 1984), are now available for mammalian cell gene mutation assays. However, while many of these systems are considered to be genetically valid, in practice only four cell lines (V79, CHO, TK6 and L5178Y) and three genetic loci (*hprt, tk* and the cell membrane Na^+/K^+ ATPase) are widely used for mutagenicity testing. In this context it is important to consider differences between the cell lines and the responses of the different loci, as discussed in the Introduction. TK6 and L5178Y cells both grow well in suspension and can be plated with high cloning efficiency in semi-solid or liquid media, two features which facilitate the treatment and plating of large numbers of cells necessary for statistical analysis. Both the *tk* locus and the *hprt* locus have been used in both cell lines (Clive *et al.*, 1979; Moore & Clive, 1982; Garner *et al.*, 1985; Evans *et al.*, 1986; Moore *et al.*, 1987a; Liber *et al.*, 1989) but the major body of data is for trifluorothymidine resistance (TFT^R) in L5178Y $TK^{+/-}$ 3.7.2c.

Various protocol variations have been developed for L5178Y and TK6, such as a fluctuation test protocol employing plating into microtitre plates (Cole *et al.*, 1986) instead of soft agar, and there are several publications describing 'modified' CHO and V79 assays where cells are allowed limited post-treatment division for mutation fixation then held at or near confluency to allow decline of pre-existing HPRT levels. The use of 12-0-tetra-decanoyl phorbol ester (TPA) to inhibit metabolic co-operation and allow higher numbers of cells to be plated has also been suggested. However, due to the multiplicity of effects of TPA including sensitisation to the toxic effects of 8-azaguanine (8AZ) (M. Fox, unpublished observations) this practice is not recommended. A forward mutation assay in which V79 or CHO cells are plated into soft agar to eliminate the effects of metabolic co-operation has also been described (Nishi *et al.*, 1984; Romert *et al.*, 1986; Oberly *et al.*, 1987). We strongly recommend further development and validation of the use of human cells (Jacobs & DeMars, 1977), which may be more appropriate for extrapolation to man, e.g. lymphocyte (untransformed) lines.

4.3.1.3 Validation

The 'validation' of an assay implies (assuming that a protocol has been developed which unequivocally detects mutant colonies of the desired type, within the definition of mutants we have considered in the Introduction) (i) that the test should be independently confirmed in several laboratories giving consistent, reproducible results and (ii) that the test should be shown to be responsive to a wide range of mutagenic chemicals from several classes, including those requiring metabolic activation and not simply to a small group of powerful direct-acting mutagens.

4.3.1.4 General technical reviews

Several recent publications are available giving guidelines for performance of assays in particular cell lines, e.g. CHO (Gupta, 1984; Li *et al.*, 1987), L5178Y $TK^{+/-}$ (Turner *et al.*, 1984; Clive *et al.*, 1987: a paper giving detailed guidelines for this assay based on the experience of many laboratories is in preparation (D. Clive *et al.*, unpublished data)), V79 (Bradley *et al.*, 1981) and TK6 human lymphoblastoid cells (Penman & Crespi, 1987). *Banbury Report* 28 (Moore *et al.*, 1987a and b) and the references therein provide a very useful overview of mutagenesis in mammalian cells.

4.3.2 Experimental design

The importance of a thorough understanding of the particular cell system in use cannot be overstressed if reproducible quantitative data are to be obtained. Some of the critical factors of general relevance are noted below, with reference to particular cell systems.

4.3.2.1 Selection of appropriate cell populations and treatment time

Statistical considerations for the numbers of cells to be treated, subcultured and plated for suspension cultures and attached cells are given in the UKEMS Guidelines Part III (Arlett *et al.*, 1989; Robinson *et al.*, 1989). To ensure that cells in all stages of the cell cycle are exposed, exponentially growing cells should be used; cells that are at or near maximum density for 24 hours prior to treatment should **not** be used. Considerable variation exists in the literature for length of treatment time, saline or medium as the vehicle and the presence or absence of serum during treatment. The presence of serum during treatment can inhibit the action of some mutagens and have no effect on others. For routine screening, a 4-hour treatment time in 5% serum is recommended. If data from different experiments are to be compared, great care should be taken to standardise treatment conditions, taking into account such factors as satisfactory growth rate of cells before treatment, and medium, serum content and cell density during treatment.

4.3.2.2 Stability of the spontaneous mutant frequency

Several workers have observed that, if cells are maintained in exponential growth for several weeks, there may be a build-up of spontaneous mutants (Thompson & Baker, 1973; Abbondandolo *et al.*, 1976; Bradley *et al.*, 1981; Fox, 1985b). For example, the L5178Y TK system seems especially subject to rapid mutant build-up (Amacher *et al.*, 1979, 1980a, b) at least for the large colony-forming mutants. Since high and variable background frequencies cause considerable problems with data interpretation (Tong & Williams, 1980), particularly when the mutagenic effect is small, every effort should be made to maintain a stable spontaneous mutant frequency. Stability may be maintained by (i) recloning to establish a line with a low spontaneous mutant frequency; a large frozen stock should then be established, and one vial used for each experiment, or (ii) cells regularly used for subculturing should be diluted to low density (e.g. 10 cells per ml). This procedure may, however, eventually lead to genetic drift and the cultures should be discarded after 3–4 months' use, and renewed from the frozen stock. In both these cases (i and ii) karyotypic stability should be checked. Alternatively, (iii) medium containing aminopterin or methotrexate and thymidine, cytosine, hypoxanthine and glycine ('HAT', 'CHAT', 'THAG' or 'THMG' medium) may be used to purge the population of TK^- or $HPRT^-$ mutants, and a large population of cells frozen for future use. However, we cannot endorse the regular weekly use of such treatment of L5178Y cells advocated by Amacher *et al.* (1979, 1980; Turner *et al.*, 1984). Indeed, such treatment is unnecessary since the required reduction in pre-existing mutants can be achieved by 'cleansing' a sub-population once, some days before its use in an experiment. Excess cells from this sub-population should be discarded. Repeated exposure of the cells to methotrexate or aminopterin is of very doubtful value. The effects of these powerful inhibitors may seriously disturb nucleotide pool sizes, are known to result in gene amplification, and may induce other enzymes such as alkaline phosphatase. It should also be borne in mind that 'cleansing' a population with HAT will not necessarily remove mutants with partial enzyme activity and may increase the spontaneous mutant frequency due to preferential growth of this type of mutant if the selective drug concentration allows them to survive. Finally, methotrexate is a clastogen.

4.3.2.3 Culture conditions

Culture conditions should be well defined. All media should support adequate growth and quality control and storage conditions should be carefully checked. References should be cited and justification, with detailed description, for major departures from the properly validated methods should be given. The importance of maintaining cells in optimal growth conditions throughout the experiment cannot be overstressed, as the efficiency of the selective agent may well depend on the growth rate of the cells (see below, Section 4.4). The factors affecting optimal growth and selection of mutants have been widely discussed in the literature and are discussed in the following sections.

(a) Serum

Batches of serum should be tested for effects on cell growth rate, CE and spontaneous and induced mutation frequency, and a large batch stored frozen. This is particularly important for 8AZ selection of HPRT mutants, as serum components (e.g. hypoxanthine) may seriously affect 8AZ toxicity. For a discussion of the use of dialysed serum, see Bradley *et al.* (1981), Li *et al.* (1987), Clive *et al.* (1987) and Oberly *et al.* (1986). Liber & Thilly (1982) strongly recommend heat inactivation of horse serum when TK mutants are to be selected (e.g. in L5178Y $TK^{+/-}$ cells and TK6 human lymphoblastoid cells) in order to eliminate a factor which degrades trifluorothymidine. Heat inactivation of serum should be at 57 °C for 30 minutes. The temperature should be carefully controlled, as too high a temperature causes complete loss of cloning efficiency. Serum should be aliquoted and stored in the dark at −20 °C and not frozen and thawed repeatedly.

(b) Medium

Medium supplemented with glutamine and sodium pyruvate can be stored for short periods in the dark at 4 °C. However, care should be taken to ensure that medium that has been stored does not lose its ability to support cell growth with time. Glutamine is unstable in liquid medium and readily cyclises. The cyclic form of glutamine is not absorbed by cells.

(c) Cell density

The maximum cell density during the selective period, which depends on the cell line and selective system used, should be carefully defined. Too high cell densities may lead to (i) loss of mutants due to overcrowding, (ii) incomplete toxicity of the selective agents, (iii) metabolic co-operation in monolayer cell lines which results in loss of mutant expression at high density.

(d) Choice and concentration of selective agent

The concentration of the selective agent should be high enough to kill all non-mutant cells. Selective agents such as trifluorothymidine which are light-sensitive and have a short half-life in medium at 37 °C should be handled with great care. Stock solutions should be stored in the dark at −20 °C, thawed just before use and any excess discarded. Both 6-thioguanine and 8-azaguanine are stable in medium plus serum for at least 48 hours, which is in excess of the time required to kill most hamster cells (Fox & Hodgkiss, 1981). For a discussion of the use of 6TG versus 8AZ for the selection of HPRT mutants in Chinese hamster cells, see for example Thacker *et al.* (1976), Bonatti *et al.* (1980), Bradley *et al.* (1981), Fox (1981), Fox & Hodgkiss (1981) and Hsie *et al.* (1981). In general, 6TG provides a more stringent selective system, although 'true' mutants at the *hprt* locus, which have reduced or altered HPRT, can be observed in 8AZ selective medium or low levels of 6TG and not in medium containing high levels of 6TG. 6TG is the agent of choice for mouse cells; 8AZ is only partially toxic as it has a low affinity for mouse HPRT (Knaap & Simmons, 1975; Cole & Arlett, 1976).

(e) Expression time

The time after treatment when the maximum frequency of newly induced mutants can be detected depends on the cell line, the selective system and the incubation conditions (see Section 4.3.1.1). The optimal time for mutant detection for a particular locus and cell line must therefore be carefully defined using several reference mutagens. In general, one expression time is adequate, e.g. 2 days for TK mutants and 6 days for selection of HPRT mutants. However, the assay measures induced mutant frequency accurately only if mutant growth rate during the expression time and cloning efficiency (CE) at the time of selection is approximately the same as that of non-mutant cells. If either or both parameters are markedly decreased in the treated population the mutagenic potency of a test chemical may be erroneously estimated. Thus, if a test chemical causes considerable growth inhibition and the only positive effect is seen at high levels of toxicity, the experiment should be extended until growth rate and CE are approximately normal. This is particularly important for assays in which a 48 hour expression time is routinely used as cells may not have fully recovered from the toxic effects of high dose treatment by this time.

In the particular case of cells in which metabolic co-operation occurs (it does not occur in mouse lymphoma cells) it is recommended that the 'respreading technique' be used at the end of the expression period (Hsie *et al.*, 1979; Fox, 1981). An alternative method involving cloning of CHO cells into medium containing agar has been described by Oberly *et al.* (1987).

(f) Agar

The importance of agar quality for optimal cloning efficiency and selection of small colony TK mutants in L5178Y $TK^{+/-}$ has been noted by Meyer *et al.* (1986).

(g) Other factors

Finally, medium, pH, temperature, humidity and cell dispersion techniques are other critical factors in mammalian cell culture experiments and should be carefully controlled if reproducible mutant frequency data are to be obtained. Low pH (<6.8) or high osmolality (>400 mOsmol/kg) may generate false positives (Brusick, 1987) particularly in the presence of S9 (Cifone *et al.*, 1987). Such effects should be measured and minimised when necessary by maintaining neutral pH and reducing the ion content of the treatment medium to compensate. However, it must be noted that pH and osmolality have been found not to affect metabolic co-operation in V79 cells (Binder & Volpenhein, 1987).

4.3.2.4 Controls and internal monitoring

(a) Tests for contamination

All cultures should be tested at regular intervals for mycoplasma contamination (Hayflick, 1973; Russel *et al.*, 1975). A mycoplasma testing service is available from suppliers of tissue culture media. Kits are available (Mycotect, BRL).

(b) Acceptable negative and positive control ranges

Positive and negative controls should yield mutant frequencies that fall within the normal range (see below for further discussion of this point). Particular care should be taken with L5178Y $TK^{+/-}$ cells to ensure that small colony TFT^R mutants are being induced and scored (D. Clive *et al.*, unpublished data).

4.3.2.5 Provision for metabolic activation

Two methods of supplying exogenous mammalian metabolic activation systems are available. (i) Uninduced, or Aroclor 1254, 3-methylcholanthrene and phenobarbitone induced S9s have been the most widely used methods, e.g. for L5178Y cells (Clive *et al.*, 1979; Amacher & Paillet, 1980; Amacher & Turner, 1982a, b; McGregor *et al.*, 1988), V79 cells (Kuroki *et al.*, 1979) and CHO cells (Carver *et al.*, 1980; Crouch *et al.*, 1979; Hsie *et al.*, 1981). (ii) Cell-mediated metabolism, using freshly isolated hepatocytes or hamster cell lines (SHE or BHK) in conjunction with V79 cells as the indicator organism (Huberman & Sachs, 1974; Newbold *et al.*, 1977; Langenbach *et al.*, 1978, 1981; Suter, 1987; Romert *et al.*, 1987). In addition to these exogenous metabolic activation systems, the use of cell types with intrinsic metabolic activity, in which activation and mutation induction occurs in the same cells, has been suggested, e.g. the adult rat epithelial cell line (ARL) studied by Tong & Williams (1980) and AHH-1, a human lymphoblastoid line (Crespi & Thilly, 1984;

Danheiser *et al.*, 1989). It is clear from the literature that no single metabolising system is ideal for all classes of chemicals. Consider, for example, the relatively simple case of the effect of S9 concentration on mutagenicity in L5178Y cells: for DMN, the higher the S9 concentration, the more active it is, while for N-acetyl-2-aminofluorene (2AAF) and B(a)P the reverse is true (Hsie *et al.*, 1979; Zeiger, 1979; Amacher & Turner, 1982b; McGregor *et al.*, 1988). Similar results have been reported by Kuroki *et al.* (1979) and Bartsch *et al.* (1979), using V79 cells.

Very variable co-factor mixes have been reported in the literature. Therefore in the Third UKEMS Collaborative Trial, a common batch of Aroclor-induced rat liver S9 was used and the co-factor constituents varied. The data showed that, provided the toxicity of co-factor constituents was carefully avoided, an NADP/G6P co-factor mix gave similar results to an NADP/isocitrate mix. It was very clear that, for B(a)P at least, 1–2% S9 for the rodent cell lines, and up to 5% S9 for the human cells, resulted in a considerably greater mutagenic response than higher S9 concentrations. Li *et al.* (1987) recommend that for CHO cells, in the first of two experiments multiple concentrations of S9 should always be used for every test substance. Thus we conclude that, while this remains a major area for further research, flexibility is important, as a single compromise protocol may not be appropriate in every case and it may be necessary to use a different protocol depending on the test substance. The laboratory conducting the test should be able to provide data showing that, using a clearly defined protocol for metabolic activation (either liver homogenate or cell-mediated), mutagens from different classes of chemicals requiring metabolic activation (e.g. B(a)P, DMN, 2AAF) induce mutations in a dose-dependent fashion. With these provisions in mind, we recommend the use of Aroclor 1254-induced S9, at concentrations of between 1 and 5% depending on the cell line, as the basis for further experiments.

4.3.2.6 Population size, replicate cultures and the numbers of cells to be sampled for mutants

(a) The numbers of cells to be treated, subcultured and plated

The investigator should be able to show that, taking toxicity and the mutant frequency in the untreated population into account, sufficient cells have been treated, subcultured and exposed to the selective agent for a significant increase over the control mean to be demonstrated. The statistical guidelines for mammalian cell gene mutation assays (Arlett *et al.*, 1989; Robinson *et al.*, 1989) contain detailed recommendations on these points. It is proposed that as a minimum requirement each treated population should contain sufficient viable cells that at least 10 spontaneous mutants will be present at the end of treatment, although others have suggested that 100 spontaneous mutants are required for statistical precision (Furth *et al.*, 1981). However, in many cases the latter suggestion is impractical. Mutant frequencies vary from $2\text{–}10 \times 10^{-5}$ for TFT resistance in L5178Y $TK^{+/-}$ cells, to $1\text{–}10 \times 10^{-6}$ for 6TG resistance and 4×10^{-7} for ouabain (OUA) resistance in some Chinese hamster cells. Thus, for the Chinese hamster cells, very large numbers of cells are needed to meet even the minimum requirement, especially when substantial kill is inflicted by the treatment. For example, if a test treatment reduced viability to 10%, a minimum of 2.5×10^{8} cells should be treated if OUA resistance is to be selected in V79 cells and statistical precision maintained. Similar large populations should be subcultured during the expression period.

The number of cells to be sampled at the end of the expression period again depends on the cell line, selective system and particular assay of choice. Specific recommendations are made in Arlett *et al.* (1989) and Robinson *et al.* (1989). The actual number of cells which can be challenged by a selective agent in a single plate is influenced by a number of factors, including metabolic co-operation and less well defined cell density effects. For plate assays, sufficient numbers of cells per plate and plates per point should be used to avoid zero counts on all plates in a set. For fluctuation assays, when the mutant frequency is estimated from multiwell trays, optimum sensitivity occurs when 50–80% of the wells are positive. This is not always possible to achieve, and eight trays for mutant selection per replicate is considered the practical limit.

A consequence of these recommendations is that, while several systems are equally valid on theoretical grounds, in practice there is a preference for suspension cultures for routine mutagenicity assays as it is easier to treat and maintain large numbers of cells, and for TFT or 6TG as the selective drug, since the frequency of OUA mutants is low in all rodent and human cells lines.

(b) Replicate cultures

Two sources of variation have been identified in experiments: first, variation in the number of cells distributed between plates from an individual culture; secondly, variation between replicate cultures undergoing the same treatment in the same experiment, and this is likely to be the larger source of variation (Leong *et al.*, 1985). Because subculturing introduces variation that cannot be estimated from a single culture

it is strongly recommended that treated and control cultures should be duplicated throughout the entire experiment.

4.3.2.7 Dose levels and cytotoxicity

The substance should first be tested for cytotoxicity under precisely the same conditions as a mutation experiment. For cells that require trypsinisation, 16–24 hours should be allowed for cell attachment before treatment commences. For the initial toxicity experiment, treatment should be over a wide range of molarity – for example, 10, 3, 1, 0.3, 0.1 mM, etc. When the toxic range has been determined, the mutation experiment should cover 3–6 doses ranging from non-toxic (90–100% survival) to toxic (i.e. *c.* 10% survival). Toxicity should be measured in the mutation assay to confirm the appropriate choice of doses. In practice, evidence of mutation induction at levels of survival less than 10% should be interpreted with great caution as sampling error at high levels of kill may lead to spurious results. A maximum concentration of 5 mg/ml of the test substance is recommended. When precipitation of the test substance occurs at relatively low toxicity, consideration should be given to a longer treatment period, e.g. 16–24 hours, at non-precipitating concentrations.

4.3.2.8 Reproducibility of experiments

The determinations should be quantitative and reproducible. Thus, the whole experiment should be done at least twice, using freshly prepared test substance, although not necessarily using the same dose range. Assuming acceptable results, if both experiments give clear-cut positive or negative results, this could be considered adequate. However, for low or equivocal mutagenic responses, especially for a small positive response (2–3 fold increase) at high toxicity in the absence of a dose response, it may be necessary to repeat the experiment more than twice if a genuine evaluation is to be made. In this case, at high levels of cytotoxicity or where there is evidence of considerable growth inhibition, the use of a longer expression time should be considered. In no circumstances can a single experiment be considered adequate.

4.3.2.9 Incubation conditions

Cells should be treated in sealed containers protected from light. Regular shaking is recommended for suspension cultures. Standard fluorescent lights in cell-handling (safety) cabinets are mutagenic (Burki & Lam, 1978) and the use of devices reducing or eliminating transmission of wavelengths below 400 nm is recommended (i.e. 'gold' fluorescent lights, or a screen of UV-absorbent material such as Perspex VE or

Plexiglas 201). The plates should be incubated under normal tissue culture conditions.

4.3.2.10 Time of scoring

This depends on the cell line and the selective agent. For the L5178Y $TK^{+/-}$ system, scoring of the 'large' and 'small' colonies (Clive *et al.*, 1987) is recommended for the positive control and for those replicates showing a significant increase over the negative control, as the additional information obtained may contribute to an understanding of the mode of action of the test compound (see Introduction 4.1.1 above).

4.4 DATA PROCESSING AND PRESENTATION

4.4.1 Recording and presentation of data

All original data, including toxicity data, absolute CE of control and treated cultures, etc., should be recorded and presented for evaluation; dose-response data from a minimum of three concentrations should be presented. The use of a single expression time may be one of the reasons for odd-shaped dose-response curves when the test compound has caused considerable growth inhibition. In such cases, data from one of the experiments should include a later expression time to ensure that adequate time has been allowed for the newly induced mutant phenotype to be expressed. If an unusual shape of dose-response curve is obtained, evidence should be given to indicate that it is not due to incomplete expression at high doses and/or reduced viability of mutants at other doses.

If a weak mutagen is under study it is particularly important to establish a dose-response relationship before making a decision as to whether it is positive or negative. Unusual dose-response relationships require a great deal of data before they can be accepted as real, particularly for weak mutagenic responses up to five times above background.

Raw mutant counts and mean number of mutant and surviving colonies per plate for control and treated cultures, and the standard deviation (SD) should be presented. Data should be tabulated, giving CE as percentage of control. Mutant frequency should be presented as mutants per 10^6 clonable cells. Possible toxicity of the vehicle should be indicated. Growth rates during the expression period may be reported but do not have to be taken into account in the calculated mutant frequency. Coulter or haemocytometer counts of the treated population during the expression period represent counts of a heterogenous cell population which includes viable cells dividing at the normal rate, cells which are dividing slowly as a result of treatment, 'dead' cells no longer capable of division, and cells which will undergo one or more divisions before ceasing to grow. Thus, 'growth rate' based on only one or two counts at 24 hour intervals is relatively meaningless even for comparative purposes. Repeated counts over at least 7 days followed by back extrapolation from the exponential growth phase may allow comparison of growth inhibition and toxicity (Cole & Arlett, 1978) but using such data to calculate mutagenic potency (Clive *et al.*, 1979) is not justified.

All validation data should be published or open to inspection.

Minimum data to be presented will depend on the activity of the mutagen and locus assayed. For strongly positive mutagens, i.e. those producing a 10–100-fold (or higher) increase above background at the highest concentration assayed (for example a dose which will give 90% cytotoxicity), data from a minimum of two experiments should be sufficiently convincing.

4.4.2 Statistical treatment of data

This has presented considerable problems in mammalian cell assays in the past and many statistical approaches have been suggested (Garner *et al.*, 1985; Murphy *et al.*, 1988; Penman & Crespi, 1987). Some of these were outlined previously (Cole *et al.*, 1983), when it was strongly recommended that a specialist group be set up to consider the problem. This has now been completed (Arlett *et al.*, 1989; Robinson *et al.*, 1989), and recommendations have been produced which cover all aspects of experimental design (including numbers of cells to be treated, subcultured and sampled for mutant frequency, replicate cultures, accuracy of dilutions and repeat experiments) and methods of analysis of the mutant frequency data, with detailed examples of data and calculations. The recommendations cover mammalian cell assays based upon colony formation and upon the fluctuation test. Both assays are similar in principle; the difference lies only in the method of estimating the mutant frequency in each population at the end of the expression time. Thus, a different statistical approach is required to evaluate the results. The recommendations, which should be carefully studied by the experimenters, may be summarised as follows:

1. Genuinely independent replicates of each treatment set up at the start, and maintained throughout the experiment, should be used. Without these it is not possible to perform a valid test of significance.
2. Sufficient cells should be treated, subcultured and plated to perform a statistical analysis. Numbers depend on the particular cell line and selective system and precise recommendations are

made (Arlett *et al.*, 1989; Robinson *et al.*, 1989). It must be stressed that **a complete set of zero plate counts gives no information of use concerning mutagenicity,** since such results cannot be analysed statistically.

3. Detailed methods for the analysis of the data and the use of the historical control mutant frequency are given. Tests are available for comparing each treatment with the control and for analysing the results for a significant linear relationship between increased mutant frequency and increased dose.
4. All experiments should be repeated and methods are given for comparing and combining results for repeat experiments.
5. The use of arbitrary 'threshold values' (e.g. × 2) has been suggested in the past. While such a concept may be a useful additional guideline, decisions on the mutagenicity of a test substance should be based on sound statistical analysis and not on an arbitrary figure.
6. If the mutant frequency at one or more doses of test substance is significantly greater than that of the negative control, and if there is a significant dose relationship, the test substance should be considered positive for this assay.

4.4.3 Ambiguous results

Experiments should show positive and negative controls within reasonable limits; acceptability of negative control mutant frequencies is defined in Arlett *et al.* (1989) and Robinson *et al.* (1989). The results at each dose-level should be consistent, and the linear trend should be consistent between experiments. If these conditions are not met, the test must be repeated. When there is disagreement between experiments, or where the controls are outside reasonable limits, further experiments are necessary. The test should also be repeated where it is of borderline significance, or where there is an effect at one dose-level only, or a trend with no individual dose-level being significant.

Another example of a possible ambiguous observation is when the mutant frequency per viable cell is only 2–3 times the control, and no absolute increase in the number of mutants is seen. While a small increase in mutant frequency may be genuine for a given mutagen, it is also possible that mutant cells may survive to form colonies on the selective plates where the cell density is high, but non-mutants fail to form colonies on the non-selective plates where the cell density is low. As a result, the mutant frequency per viable cell is artificially high. Such small increases should be treated with caution. This is particularly important at high levels of kill when there may not be an absolute increase in the number of mutants per plate, and no dose-related increase is seen at less toxic concentrations of the test substance.

Such ambiguity may result from the use of non-stringent selective conditions particularly for TK^- and $HPRT^-$. The stringency of selection can be reduced by mutagen treatment, since this slows cellular growth rate and reduces incorporation of analogues into DNA/RNA which is necessary for killing of non-mutant cells. It is very important to ensure the stringency of selection for each particular cell line. Nucleotide pool sizes differ between cell lines and under different growth conditions. The relative affinities of the salvage enzyme for the analogue and natural substrate can also differ markedly between different cell lines and in different species. Apparent affinity of the analogue for the target enzyme can also be influenced by variations in, for example, thymidine and hypoxanthine in serum.

Some evidence of stringency of selective conditions employed in particular experiments should be provided particularly if a high or very variable background mutant frequency is obtained: for example, reproducibility of the kill curve of the selective agent for non-mutant cells on different occasions, or an indication that, in spite of variations in sera, the concentration of selective agent is sufficiently high to ensure that all non-mutant cells are killed.

4.5 CONCLUSIONS AND RECOMMENDATIONS

Discussion of how the critical factors identified above may influence the validity of the data and the interpretation of results has in many instances been included in each of the sections above. Several areas have been identified which are important if mammalian cell gene mutation assays are to be used with confidence to provide reproducible, quantitative data upon which estimates of mutagenic potency and hazard evaluation can be based.

4.5.1 Genetic systems and cell lines

The group recommends that the L5178Y $TK^{+/-}$ and TK6 systems as assays of choice, for the following reasons:

1. the ease of treatment and plating of large numbers of cells necessary for the statistical analysis of results;
2. the lack of metabolic co-operation;
3. the possibility that if both large and small colonies are scored in $TK^{+/-}$ cells both cytogenetic and point mutation events can be detected. Emphasis should possibly be placed on further work

with TK6 because as human cells they may be more relevant. However, it is important to remember that L5178Y $TK^{+/-}$, and particularly TK6 cells, are much more sensitive to the cytotoxic effects of many agents than CHO and V79 cells and may be hypersensitive to mutagenic effects even when assayed using well designed protocols.

4.5.2 Metabolic activation systems

Some progress has been made in characterising the optimal conditions for metabolic activation using various sources of S9 (McGregor *et al.*, 1988). Until the potential for using the same cells for metabolic conversion and mutant selection is further researched and validated, Aroclor-induced rat S9 remains the system of choice. Experience has shown that the use of 1–2% S9 should be satisfactory in most cases. However, it is clear that no single source or level of S9 will be optimal in every case and it is essential that a flexible approach should be used.

4.5.3 Statistical analysis

1. Experiments should be designed to maximise statistical analysis of the data. Large populations of cells should be treated, subcultured and plated in selective medium, the numbers depending on cell line and system. Genuinely independent replicates should be incorporated into every experiment. All experiments should be repeated.
2. Statistical tests are available for comparing each treatment with the control, for analysing the data for linear trend and for comparing and combining experiments. Appropriate methods are available for plate and fluctuation assays (see Arlett *et al.*, 1989, and Robinson *et al.*, 1989).
3. We strongly discourage the use of arbitary decisions (e.g. a 2- or 3-fold incease over the relevant control data) indicating positive or negative results.

4.6 REFERENCES

Abbondandolo, A., Bonatti, S., Collela, C., Corti, G., Mattencci, F., Mazzaccaro, A. & Rainaldi, G. (1976). A comparative study of different experimental protocols for mutagenesis assays with the 8-azaguanine resistance system in cultured Chinese Hamster cells. *Mutation Research*, **37**, 293–306.

Abbondandolo, A., Bonatti, S., Corsi, C., Cori, G., Fiorio, R., Leporini, C., Mazzaccaro, A., & Nieri, R. (1980). The use of organic solvents in mutagenicity testing. *Mutation Research*, **79**, 141–50.

Albertini, R.J., Castle, K.L. & Borcherding, W.R. (1982). T-cell cloning to detect the mutant 6-thioguanine resistant lymphocytes present in human peripheral blood. *Proceedings of the National Academy of Sciences (USA)*, **79,** 6617–21.

Albertini, R.J., O'Neill, J.P., Nicklas, J.A., Heintz, N.H. & Kelleher, P.C. (1985). Alterations of the hprt gene in human *in vivo*-derived 6-thioguanine-resistant T lymphocytes. *Nature,* **316,** 369–71.

Amacher, D.E. & Paillet, S.C. (1980). Promutagen activation by rodent liver post-mitochondrial fractions in the L5178Y/TK mutation assay. *Mutation Research,* **74,** 485–501.

Amacher, D.E., Paillet, S.C. & Ray, V.A. (1979). Point mutations at the thymidine kinase locus in L5178Y mouse lymphoma cells. 1. Application to genetic toxicological testing. *Mutation Research,* **64,** 391–406.

Amacher, D.E., Paillet, S.C., Turner, G.N., Ray, V.A. & Salsburg, D.S. (1980). Point mutations at the thymidine kinase locus in L5178Y mouse lymphoma cells. 2. Test validation and interpretation. *Mutation Research,* **72,** 447–74.

Amacher, D.E. & Turner, G.N. (1982a). Mutagenic evaluation of carcinogens and non-carcinogens in the L5178Y/TK assay using post-mitochondrial fractions (S9) from normal rat liver. *Mutation Research,* **97,** 49–65.

Amacher, D.E. & Turner, G.N. (1982b). The effect of liver post-mitochondrial fraction concentration from aroclor-1254 treated rats on pro-mutagen activation in L5178Y cells. *Mutation Research,* **97,** 131–7.

Applegate, M.L. & Hozier, J.C. (1987). On the complexity of mutagenic events at the mouse lymphoma thymidine kinase locus. In *Banbury Report 28,* ed. M.M. Moore, D.M. Demarini, F.J. de Serres and K.R. Tindall. Cold Spring Harbor Laboratories, New York, pp. 213–24.

Arlett, C.F., Smith, D.M., Clark, G.M., Green, M.H.L., Cole, J., McGregor, D.B. & Asquith, J.C. (1989). Mammalian cell assays based upon colony formation. In *UKEMS Sub-committee on Guidelines for Mutagenicity Testing. Report. Part III. Statistical Evaluation of Mutagenicity Test Data,* ed. D.J. Kirkland. Cambridge University Press, Cambridge, pp. 66–101.

Baker, R.M., Brunette, D.M., Mankovitz, R., Thompson, L.H., Whitmore, G.F., Siminovitch, L. & Till, J.E. (1974). Ouabain resistant mutants of mouse and Chinese hamster cells. *Cell,* **1,** 9–21.

Barbacid, M. (1986). Mutagens, oncogenes and cancer. *Trends in Genetics,* **2,** 188–92.

Bartsch, C., Malaveille, A.M., Camus, G., Martel-Planche, G., Brun, A., Hautefeuille, N., Sabadie, A., Barbin, T., Kuroki, C., Drevon, C. & Montesano, R. (1979). Validation and comparative studies on 180 chemicals with *S. typhimurium* strains and V79 Chinese hamster cells in the presence of various metabolising systems. *Mutation Research,* **76,** 1–50.

Binder, R.L. & Volpenhein, M.E. (1987). An evaluation of the effects of culture medium osmolality and pH on metabolic co-operation between Chinese hamster V79 cells. *Carcinogenesis,* **8,** 1257–61.

Bonatti, S., Abbondandalo, A., Mazzaccaro, A., Biorio, R. & Mariani, L. (1980). Experiments on the effect of medium in mutation tests with the HPRT system in cultured cells. *Mutation Research,* **72,** 475–82.

Bradley, M.O., Bhuyan, B., Francis, M.C., Langenbach, R., Peterson, A. & Huberman, E. (1981). Mutagenesis by chemical agents in V79 Chinese hamster cells: a review and analysis of the literature. A report of the genetox programme. *Mutation Research,* **87,** 81–142.

Bradley, W.E.C., Gareau, J.L.P., Seifert, A.M. & Messing, K. (1987).

Molecular characterisation of 15 rearrangements among 90 human *in vivo* somatic mutants shows that deletions predominate. *Molecular and Cellular Biology*, **7**, 956–60.

Breimer, L.H., Nalbantoglu, J. & Meuth, M. (1986). Structure and sequence of mutations induced by ionising radiation at selectable loci in Chinese hamster ovary cells. *Journal of Molecular Biology*, **192**, 669–74.

Brown, R. & Thacker, J. (1984). The nature of mutants induced by ionising radiation in cultured hamster cells. I. Isolation and initial characterization of spontaneous ionising radiation induced and ethylmethanesulphonate induced mutants resistant to 6-thioguanine. *Mutation Research*, **129**, 269–81.

Brown, R., Stretch, A. & Thacker, J. (1986). The nature of mutants induced by ionising radiation in cultured hamster cells. II. Antigenic response and reverse mutation of HPRT-deficient mutants induced by gamma-rays or ethylmethanesulphonate. *Mutation Research*, **160**, 111–20.

Brusick, D.J. (1987). Implications of treatment condition-induced-genotoxicity for chemical screening and data interpretation. *Mutation Research*, **189**, 1–6.

Burki, H.J. & Lam, C.K. (1978). Comparison of the lethal and mutagenic effects of gold and white fluorescent lights on cultured mammalian cells. *Mutation Research*, **54**, 373–7.

Carver, J.H., Adair, G.M. & Wandres, D.L. (1980). Mutagenicity testing in mammalian cells. 2. Validation of multiple drug-resistance markers having practical application for screening potential mutagens. *Mutation Research*, **72**, 207–30.

Cifone, M.A., Myhr, B., Eiche, A. & Bolcsfoldi, G. (1987). Effect of pH shifts on the mutant frequency at the thymidine kinase locus in mouse lymphoma L5178Y $TK^{+/-}$ cells. *Mutation Research*, **189**, 39–46.

Clive, D., Caspary, W., Kirkby, P.E., Krehl, R., Moore, M., Mayo, J. & Oberly, T.J. (1987). Guide for performing the mouse lymphoma assay for mammalian cell mutagenicity. *Mutation Research*, **189**, 143–56.

Clive, D., Johnson, K.O., Spector, J.F.S., Batson, A.G. & Brown, M.M.M. (1979). Validation and characterisation of the L5178Y $TK^{+/-}$ mouse lymphoma mutagen assay system. *Mutation Research*, **59**, 61–108.

Cole, J. & Arlett, C.F. (1976). Ethylmethanesulphonate mutagenesis with L5178Y mouse lymphoma cells: a comparison of ouabain, thioguanine and excess thymidine resistance. *Mutation Research*, **34**, 507–26.

Cole, J. & Arlett, C.F. (1978). Methyl methane sulphonate mutagenesis in L5178Y mouse lymphoma cells. *Mutation Research*, **50**, 111–20.

Cole, J., Fox, M., Garner, C., McGregor, D. & Thacker, J. (1983). Gene mutation assays in mammalian cells. In *UKEMS Sub-committee on Guidelines for Mutagenicity Testing. Report. Part I. Basic Test Battery*, ed. B.J. Dean. United Kingdom Environmental Mutagen Society, Swansea, pp. 65–102.

Cole, J., Green, M.H.L., James, S.E., Henderson, L. & Cole, H. (1988). A further assessment of factors influencing measurements of thioguanine-resistant mutant frequency in circulating T lymphocytes. *Mutation Research*, **204**, 493–507.

Cole, J., Muriel, W.J. & Bridges, B.A. (1986). The mutagenicity of sodium fluoride to L5178Y (wild-type and $TK^{+/-}$ 3.7.2c) mouse lymphoma cells. *Mutagenesis*, **1**, 157–67.

Crespi, C.L. & Thilly, W.G. (1984). Assay for gene mutation in a human lymphoblast line, AHH-1, competent for xenobiotic metabolism. *Mutation Research*, **128**, 221–30.

Crouch, D.B., Bermudez, E., Decad, G.M. & Dent, J.G. (1979). The influence of activation systems on the metabolism of 2,4-dinitrotoluene and its mutagenicity in CHO cells. In *Banbury Report 2,* ed. A.W. Hsie, J.P. O'Neill and V.K. McElroy. Cold Spring Harbor Laboratory, New York, pp. 303–9.

Danheiser, S.L., Liber, H.L. & Thilly, W.G. (1989). Long-term, low-dose benzo(a)-pyrene-induced mutation in human lymphoblasts competent in xenobiotic metabolism. *Mutation Research,* **210,** 143–7.

DeMarini, D.M., Brockman, H.E., de Serres, F.J., Evans, H.H., Stankowski, L.F. Jr & Hsie, A.W. (1989). Specific-locus mutations induced in eukaryotes (especially mammalian cells) by radiation and chemicals: a perspective. *Mutation Research,* **220,** 11–29.

DuBridge, R.B. & Calos, M.P. (1987). Molecular approaches to the study of gene mutation in human cells. *Trends in Genetics,* **3,** 293–7.

Dubrovic, A., Gareau, J.L.P., Ouellette, G. & Bradley, W.E.C. (1988). DNA methylation and gene inactivation at thymidine kinase locus: two different mechanisms for silencing autosomal genes. *Somatic Cellular and Molecular Genetics,* **14,** 55–68.

Evans, H.H., Mencl, J., Horng, M.F., Ricanti, M., Sanchez, C. & Hozier, J. (1986). Locus specificity in the mutability of L5178Y mouse lymphoma cells: the role of multilocus lesions. *Proceedings of the National Academy of Sciences (USA),* **83,** 4379–85.

Fox, M. (1981). Some quantitative aspects of the response of mammalian cells in vitro to induced mutagenesis. In *Cancer Biology Reviews vol. 3,* (ed. J.J. Marchelonis and M.G. Hanna). Marcel Decker Inc., New York and Basel, pp. 23–62.

Fox, M. (1985a). Assay of colony forming ability in established cell lines. In *Cell Clones,* ed. C.S. Potten and J.H. Hendry. Churchill Livingstone. pp. 185–95.

Fox, M. (1985b). The effects of pyrimidine nucleotides on alkylating agent induced cytotoxicity and spontaneous and induced mutation to purine analogue resistance in V79 cells. In *Genetic Consequences of Nucleotide Pool Imbalance,* ed. F.J. de Serres. Plenum, New York, pp. 435–53.

Fox, M. & Hodgkiss, R.J. (1981). Mechanisms of cytotoxic action of azaguanine and thioguanine in wild type V79 cell lines and their relative efficiency in selection of structural gene mutants. *Mutation Research,* **80,** 165–85

Furth, E.E., Thilly, W.G., Penman, B.W., Liber, H.L. & Rand, W.M. (1981). Quantitative assay for mutation in diploid human lymphoblasts using microtiter plates. *Analytical Biochemistry,* **110,** 1–8.

Fuscoe, J.C., Ockey, C.H. & Fox, M. (1986). Molecular analysis of X-ray induced mutants at the HPRT locus in V79 Chinese hamster cells. *International Journal of Radiation Biology,* **49,** 1011–20.

Garner, R.C., Amacher, D., Caspary, W., Crespi, C., Delow, D., Knaap, A., Kuroda, Y., Kuroki, T., Myhr, B., Oberly, T., Styles, J. & Zdzienicka, M. (1985). Summary report on the performance of gene mutation assays in mammalian cells in culture. In *Evaluation of Short-term Tests for Carcinogens. Progress in Mutation Research, Vol. 5,* ed. J. Ashby, F.J. de Serres, M. Draper, M. Ishidate, B.H. Margolin, B.E. Matter and M.D. Shelby. Elsevier, New York, pp. 85–94.

Gebara, M.M., Drevon, C., Harcourt, S.A., Steingrimsdottir, H., James, M.R., Burke, J.F., Arlett, C.F. & Lehmann, A.R. (1987). Inactivation of a transfected gene in human fibroblasts can occur by deletion, amplification, phenotypic switching, or methylation. *Molecular and Cellular Biology,* **7,** 1459–64.

Gupta, R.S. (1984). Genetic markers for quantitative mutagenesis studies in CHO cells: Applications to mutagen screening. In *Handbook of Mutagenicity Test Procedures, 2nd ed*, ed. B.J. Kilbey, M. Legator, W. Nichols and C. Ramel. Elsevier, Amsterdam, pp. 291–319.

Hayflick, L. (1973). Screening cultures for mycoplasma infections. In *Tissue Culture Methods and Applications*, ed. P.F. Kruse and M.K. Petterson. Academic Press, New York, pp. 722–8.

Henderson, L., Cole, H., Cole, J., James, S.E. & Green, M. (1986). Detection of somatic mutations in man: evaluation of the microtitre cloning assay for T-lymphocytes. *Mutagenesis*, **1**, 195–200.

Hozier, J., Sawyer, J., Clive, D. & Moore, M.M. (1985). Chromosome 11 aberrations in small colony L5178Y $TK^{+/-}$ mutants early in their clonal history. *Mutation Research*, **147**, 237–42.

Hozier, J., Sawyer, J., Moore, M., Howard, B. & Clive, D. (1981). Cytogenetic analysis of the L5178Y/$TK^{+/-}$ to $TK^{-/-}$ mouse lymphoma mutagenesis assay system. *Mutation Research*, **84**, 169–81.

Hsie, A.W. (1987). The use of the *hgprt* versus *gpt* locus for quantitative mammalian cell mutagenesis. In *Banbury Report 28*, ed. M. Moore, D.M. DeMarini, F.J. de Serres and K.R. Tindall. Cold Spring Harbor Laboratory, New York, pp. 37–46.

Hsie, A.W., Casciano, D.A., Crouch, D.B., Krahn, D.F., O'Neill, J.P. & Whitfield, B.L. (1981). The use of Chinese hamster ovary cells to quantify specific locus mutation and to determine mutagenicity of chemicals. A report of the genetox programme. *Mutation Research*, **86**, 193–214.

Hsie, A.W., O'Neill, J.P. & McElheny, B.K. (ed.) (1979). *Banbury Report 2: Mammalian Cell Mutagenesis: The maturation of test systems*. Cold Spring Harbor Laboratory, New York.

Huberman, E. & Sachs, L. (1974). Cell mediated mutagenesis of mammalian cells with chemical carcinogens. *International Journal of Cancer*, **13**, 326–33.

Jacobs, L. & DeMars, R. (1977). Chemical mutagenesis with diploid human fibroblasts. In *Handbook of Mutagenicity Test Procedures*, ed. B.J. Kilbey, M. Legator, W. Nichols and C. Ramel. Elsevier, Amsterdam, pp. 193–220.

Jensen, J.C. & Thilly, W.G. (1986). Spontaneous and induced chromosomal aberrations and gene mutations in human lymphoblasts: Mitomycin C methylnitrosourea and ethylnitrosourea. *Mutation Research*, **160**, 95–102.

Knaap, A.G.A.C. & Simmons, J.W.I.M. (1975). A mutational assay system for L5178Y mouse lymphoma cells using hypoxanthine guanine phosphoribosyl transferase deficiency as a marker. The occurrence of a long expression time for mutations induced by X-rays and EMS. *Mutation Research*, **30**, 97–110.

Kuroki, T., Malaveille, C., Drevon, C., Piccoli, C., MacLeod, M. & Selkirk, J.K. (1979). Critical importance of microsome concentration in mutagenesis assays with V79 Chinese hamster cells. *Mutation Research*, **63**, 259–72.

Langenbach, R., Freed, H.J. & Huberman, E. (1978). Liver cell-mediated mutagenesis of mammalian cells by liver carcinogens. *Proceedings of the National Academy of Sciences (USA)*, **75**, 2864–7.

Langenbach, R., Nesnow, S., Tompa, A., Gingell, R. & Kuszynski, C. (1981). Lung and liver cell mediated mutagenesis systems: Specificities in activation of chemical carcinogens. *Carcinogenesis*, **2**, 851–8.

Langlois, R.G., Bigbee, W.L., Kyoizumi, S., Nakamura, N., Bean, M.A., Akiyama, M. & Jensen, R.H. (1987). Evidence for increased somatic cell mutations at the glycophorin A locus in atomic bomb survivors. *Science*, **236**, 445–8.

Leong, P.-M., Thilly, W.G. & Morgenthaler, S. (1985). Variance estimation in single-cell mutation assays: Comparison to experimental observations in human lymphoblasts at 4 gene loci. *Mutation Research*, **150**, 403–10.

Li, A.P., Carver, J.H., Choy, W.N., Gupta, R.S., Loveday, K.S., O'Neill, J.P., Riddle, J.C., Stankowski, L.F. & Yang, L.L. (1987). A guide for the performance of the Chinese Hamster ovary cell/hypoxanthine guanine phosphoribosyl transferase gene mutation assay. *Mutation Research*, **189**, 135–41.

Liber, H.L., Leong, P.M., Terry, V.H. & Little, J.B. (1986). X-rays mutate human lymphoblast cells at genetic loci that should only respond to point mutations. *Mutation Research*, **163**, 91–7.

Liber, H.L. & Thilly, W.G. (1982). Mutation assay at the thymidine kinase locus in human diploid lymphoblasts. *Mutation Research*, **94**, 467–85.

Liber, H.L., Yandell, D.W. & Little, J.B. (1989). A comparison of mutation induction at the *tk* and *hprt* loci in human lymphoblastoid cells; quantitative differences are due to an additional class of mutations at the autosomal *tk* locus. *Mutation Research*, **216**, 9–17.

McGregor, D.B., Edwards, I., Riach, C.G., Cattenach, P., Martin, R., Mitchell, A. & Caspary, W.J. (1988). Studies of an S9 based metabolic activation system used in the mouse lymphoma L5178Y cell mutation assay. *Mutagenesis*, **3**, 485–90.

McKenna, P.G. & Ward, P.E. (1987). Mutation at the APRT locus in Friend Erythroleukaemia cells. 1. Mutation rates and properties of mutants. *Mutation Research*, **180**, 267–71.

Melton, D.W., Konecki, D.S., Brennand, J. & Caskey, C.T. (1984). Structure, expression, and mutation of the hypoxanthine phosphoribosyltransferase gene. *Proceedings of the National Academy of Sciences (USA)*, **81**, 2147–51.

Meyer, M., Brock, K., Lawrence, K., Casto, B. & Moore, M.M. (1986). Evaluation of the effect of agar on the results obtained in the L5178Y mouse lymphoma assay. *Environmental Mutagenesis*, **8**, 727–40.

Moore, M.M., Applegate, M.L. & Hozier, J.C. (1988). Do trifluorothymidine-resistant mutants of L5178Y mouse lymphoma cells re-express thymidine kinase activity following 5-azacytidine treatment? *Mutation Research*, **207**, 77–82.

Moore, M.M., Brock, K.H., DeMarini, D.M. & Doerr, C.L. (1987a). Differential recovery of induced mutants at the *tk* and *hgprt* loci in mammalian cells. In *Banbury Report 28*, ed. M.M. Moore, D.M. DeMarini, F.J. de Serres & K.R. Tindall. Cold Spring Harbor Laboratory, New York, pp. 93–108.

Moore, M.M. & Clive, D. (1982). The quantitation of $Tk^{+/-}$ and HGPRT mutants of L5178Y $TK^{+/-}$ mouse lymphoma cells at varying times post-treatment. *Environmental Mutagenesis*, **4**, 499–519.

Moore, M.M., Clive, D., Howard, B.E., Batson, A.G. & Turner, N.T. (1985a). *In situ* analysis of trifluorothymidine-resistant (TFT^R) mutants of L5178Y $TK^{+/-}$ mouse lymphoma cells. *Mutation Research*, **151**, 147–59.

Moore, M.M., Clive, D., Hozier, J.C., Howard, B.E., Batson, A.G., Turner, N.T. & Sawyer, J. (1985b). Analysis of trifluorothymidine-resistant mutants of L5178Y/$TK^{+/-}$ mouse lymphoma cell. *Mutation Research*, **151**, 161–74.

Moore, M.M., DeMarini, D.M., de Serres, F.J. & Tindall, K.R. (ed.) (1987b). *Banbury Report 28: Mammalian Cell Mutagenesis*. Cold Spring Harbor Laboratory, New York.

Morley, A.A., Trainor, K.J., Dempsey, J.L. & Seshadri, R.S. (1985). Methods for study of mutations and mutagenesis in human lymphocytes. *Mutation Research*, **147**, 363–7.

Muriel, W.J., Cole, J. & Lehmann, A.R. (1987). Molecular analysis of ouabain-resistant mutants of the mouse lymphoma cell line L5178Y. *Mutagenesis*, **1**, 383–9.

Murphy, S.A., Caspary, W.J. & Margolin, B.H. (1988). A statistical analysis for the mouse lymphoma cell forward mutation assay. *Mutation Research*, **303**, 145–54.

Myhr, B.C. & Mayo, J.K. (1987). Mutagenicity of rat liver S9 to L5178Y mouse lymphoma cells. *Mutation Research*, **189**, 27–37.

Newbold, R.F., Wigley, C.B., Thompson, M.H. & Brookes, P. (1977). Cell mediated mutagenesis in cultured Chinese hamster cells by carcinogenic hydrocarbons: Nature and extent of the associated hydrocarbon DNA reaction. *Mutation Research*, **43**, 101–16.

Nishi, Y., Hasegana, M.M. & Inui, N. (1984). Forward mutation assay of V79 cells to 6-thioguanine resistance in a soft agar technique that eliminates effects of metabolic co-operation. *Mutation Research*, **125**, 105–14.

Oberly, T.J., Bewsey, B.J. & Probst, G.S. (1986). Thymidine kinase activity and trifluorothymidine resistance of spontaneous and mutagen-induced L5178Y cells in RPMI 1640 medium. *Mutation Research*, **161**, 165–71.

Oberly, T.J., Bewsey, B.J. & Probst, G.S. (1987). A procedure for the CHO/HGPRT mutation assay involving treatment of cells in suspension culture and selection of mutants in soft agar. *Mutation Research*, **182**, 99–111.

Paeratakul, U. & Taylor, M.W. (1986). Isolation and characterization of mutants at the APRT locus in the L5178Y $TK^{+/-}$ mouse lymphoma cell line. *Mutation Research*, **160**, 61–9.

Penman, B.W. & Crespi, C.L. (1987). Analysis of human lymphoblast mutation assays by using historical negative control data bases. *Environmental and Molecular Mutagenesis*, **10**, 35–60.

Robinson, W.D., Healy, M.J.R., Green, M.H.L., Cole, J., Gatehouse, D. & Garner, R.C. (1989). Statistical evaluation of bacterial/mammalian fluctuation tests. In *UKEMS Sub-committee on Guidelines for Mutagenicity Testing. Report. Part III. Statistical Evaluation of Mutagenicity Test Data*, ed. D.J. Kirkland. Cambridge University Press, Cambridge, pp. 102–40.

Romert, L., Andesson, O., Zhang, L.-H., Ripe, E. & Jenssen, D. (1987). Mutagenicity studies by co-cultivation of bronchoalveolar cells and blood lymphocytes with V79 Chinese hamster cells. *Mutation Research*, **178**, 123–34.

Romert, L., Zhang, L.H. & Jenssen, D. (1986). A method with enhanced sensitivity for the induction of 6TG resistant mutants in V79 Chinese hamster cells. *Mutation Research*, **175**, 103–6.

Russel, W.C., Newman, C. & Williamson, D.H. (1975). A simple cytochemical technique for demonstration of DNA in cells infected with mycoplasmas and viruses. *Nature*, **253**, 461–2.

Stankowski, L.F., Tindall, K.R. & Hsie, A.W. (1986). Quantitative and molecular analyses of ethyl methanesulphonate and ICR191-induced mutation in AS52 cells. *Mutation Research*, **160**, 133–47.

Steglich, C. & DeMars, R. (1982). Mutations causing deficiency of APRT in fibroblasts cultured from humans heterozygous for mutant APRT alleles. *Somatic Cellular and Molecular Genetics*, **8**, 115–41.

Suter, W. (1987). Mutagenicity of procarbazine for V79 Chinese hamster fibroblasts in the presence of trans metabolic activation systems. *Mutagenesis*, **2**, 27–32.

Thacker, J. (1985). The molecular nature of mutations in cultured mammalian cells: a review. *Mutation Research*, **150**, 431–42.

Thacker, J. & Ganesh, A.N. (1989). Molecular analysis of spontaneous and ethyl methane sulphonate-induced mutations of the hprt gene in hamster cells. *Mutation Research*, **210**, 103–12.

Thacker, J., Stephans, M.A. & Stretch, A. (1976). Factors affecting the efficiency of purine analogues as selective agents for mutants of mammalian cells induced by ionising radiation. *Mutation Research*, **35**, 465–78.

Thompson, L.H. & Baker, R.M. (1973). Isolation of mutants of cultured mammalian cells. In *Methods in Cell Biology IV*, ed. D.M. Prescot. Academic Press, New York, pp. 209–81.

Thompson, L.H., Fong, S. & Brookman, K. (1980). Validation of conditions for efficient detection of HPRT and APRT mutations in suspension-cultured Chinese hamster ovary cells. *Mutation Research*, **74**, 21–36.

Tong, C. & Williams, G.M. (1980). Definition of conditions for the detection of genotoxic chemicals in adult rat-liver epithelial cell/hypoxanthine guanine phosphoribosyl transferase (ARL/HGPRT) mutagenesis assay. *Mutation Research*, **74**, 1–9.

Turner, N.T., Batson, A.G. & Clive, D. (1984). Procedures for the L5178Y $TK^{+/-} - TK^{-/-}$ mouse lymphoma assay. In *Handbook of Mutagenicity Test Procedures*, ed. B.J. Kilbey, M. Legator, W. Nichols and C. Ramel. Elsevier, Amsterdam, pp. 239–68.

Vrieling, H., Simons, J.W.I.M., Arwert, F., Natarajan, A.T. & Van Zeeland, A.A. (1985). Mutations induced by X-rays at the HPRT locus in cultured Chinese hamster cells are mostly large deletions. *Mutation Research*, **144**, 281–6.

Vrieling, H., Niericker, M.J., Simons, J.W.I.M. & van Zeeland, A.A. (1988). Molecular analysis of mutations induced by N-ethyl-N-nitrosourea at the HPRT locus in mouse lymphoma cells. *Mutation Research*, **198**, 99–106.

Waldren, C., Jones, C. & Puck, T.T. (1979). Measurement of mutagenesis in mammalian cells. *Proceedings of the National Academy of Sciences (USA)*, **76**, 1358–62.

Ward, P.E. & McKenna, P.G. (1987). Mutation at the APRT locus in Friend erythroleukaemia cells. 2. Sensitivity to mitomycin C induced cytogenetic damage. *Mutation Research*, **180**, 273–6.

Yandell, D.W., Dryja, T.P. & Little, J.B. (1986). Somatic mutations at a heterozygous autosomal locus in human cells occur more frequently by allele loss than by intragenic structural alterations. *Somatic Cellular and Molecular Genetics*, **12**, 255–63.

Yang, T.P., Patel, P.I., Chinault, A.C., Stout, J.T., Jackson, L.G., Hildebrand, B.M. & Caskey, C.T. (1984). Molecular evidence for new mutation at the hprt locus in Lesch-Nyhan patients. *Nature*, **310**, 412–4.

Zeiger, E. (1979). Round table identifying activation systems. In *Mammalian Cell Mutagenesis: The Maturation of Test Systems, Banbury Report 2*, ed. A.W. Hsie, J.P. O'Neill and V.K. McElheny. Cold Spring Harbor Laboratory, New York, pp. 319–25.

5

In vivo cytogenetics assays

M. RICHOLD A. CHANDLEY
J. ASHBY D.G. GATEHOUSE
J. BOOTMAN L. HENDERSON

5.1 INTRODUCTION

5.1.1 Principles and genetic basis

In vivo cytogenetic assays detect chromosome damage induced in whole animals following administration of test compounds. The tests can be performed in somatic or germinal cells using either direct analysis of chromosome damage (metaphase analysis) or indirect parameters (micronucleus formation). The gross chromosome damage detected in these assays is frequently lethal to the cell. However, its presence indicates a potential to induce more subtle chromosome damage than is compatible with cell division and/or to induce similar damage in somatic cells or in germinal cells which may lead to heritable cytogenetic abnormalities. Cytogenetic damage is also indicative of the interaction of a test compound with DNA and consequently of its potential to induce other genotoxic damage (e.g. gene mutation). Apart from detecting chromosome breakage events the micronucleus test is capable of detecting chemicals which induce whole chromosome loss (aneuploidy) in the absence of clastogenic activity (Tsuchimoto & Matter, 1979). The micronucleus test is generally comparable in sensitivity to chromosome analysis (Tsuchimoto & Matter, 1979; Kliesch *et al.*, 1981; Ashby *et al.*, 1988), yet is simpler and cheaper to perform.

The importance of chromosome changes in the aetiology of neoplasia has been highlighted in recent years by advances in the field of oncogene research. Whilst it has been known for many years that some human neoplasms are associated with specific chromosome changes (reviewed in Yunis, 1983), it is now known that chromosome rearrangements may induce neoplasia by activating a proto-oncogene, as in Burkitt's lymphoma (Adams *et al.*, 1983), or by producing a novel DNA transcript such as is found in chronic myeloid leukaemia, where a translocated oncogene and the adjacent gene are transcribed to form a new RNA (Shtivelman *et al.*, 1985). Various types of cytogenetic events can lead to neoplasia in humans, e.g. reciprocal translocations, deletions or non-reciprocal rearrangements resulting in loss of chromatin and duplication of whole chromosomes or chromosome segments (reviewed in Gilbert, 1983).

This chapter discusses the conduct of cytogenetic assays which measure gross chromosome changes in somatic or germ cells. It does not include the *in vivo* sister chromatid exchange assay which was covered in Part II of the UKEMS Guidelines (Topham *et al.*, 1983).

5.1.2 Types of assay

5.1.2.1 Somatic cell assays

(a) Micronuclei

A more simple method of measuring chromosome damage than metaphase analysis is the assessment of induced micronuclei in proliferating cells. The cell type most widely used is the polychromatic erythrocyte (PCE) in bone marrow. Whilst the mouse is the most frequently used species, others such as the rat are also considered appropriate. The types of cytogenetic effects detected by this technique are chromosome breakage resulting in centric and acentric fragments and numerical changes due to chromosome loss during cell division. Such damage, induced in the immature erythroblast, results in the production of a micronucleus outside the main nucleus. The micronucleus may not be extruded with the nucleus when the erythroblast matures to form the polychromatic erythrocyte and is easily detectable as a chromatin containing body in the enucleated cell.

Micronuclei can also be analysed in polychromatic erythrocytes in peripheral blood (MacGregor *et al.*, 1980), in liver hepatocytes following hepatectomy (Tates *et al.*, 1980) or after stimulation with 4-acetylaminofluorene (Braithwaite & Ashby, 1988), or, in principle, in any tissue containing cells which are proliferating or which can be induced to proliferate and which have an adequate cytoplasmic to nuclear ratio.

(b) Metaphase analysis

In principle, metaphase analysis can be performed in any tissue containing dividing cells. Whilst the bone marrow is the most appropriate tissue for screening purposes, other cells may be examined when tissue specific effects are of interest. Following treatment with a test compound, dividing cells are arrested in metaphase by the administration of colchicine or Colcemid at various intervals after treatment. Preparations are analysed for structural chromosomal damage. Because bone marrow cells are rapidly proliferating they are sensitive to both S-dependent and S-independent mutagens. The bone marrow is well perfused with blood vessels and therefore the cells should be exposed to test compound or its long-lived metabolites present in the peripheral circulation. However, the metabolic capability of the bone marrow is lower than that of some other tissues, particularly liver, and therefore this tissue may be insensitive to the genotoxic effects of some chemicals which are activated in other tissues.

Chromosome damage may also be assessed in peripheral blood lymphocytes even though the target cell is likely to be insensitive (Newton & Lilly, 1986). The proportion of lymphocytes in peripheral blood which are dividing is very low so it is necessary to stimulate the cells to divide *in vitro* by the use of mitogens. The lymphocytes are in G_0 during exposure *in vivo* and will be relatively insensitive to S-dependent mutagens, and any induced damage may be repaired *in vivo* prior to culture or *in vitro* in the time between culture initiation and harvesting.

5.1.2.2 Germinal cell chromosome assays

An obvious advantage of examining germinal cells is that the tissue examined is concerned with gamete production and thus potentially highly relevant to an assessment of heritable cytogenetic damage. However, it is apparent from work carried out to date that the list of chemicals known to be mutagenic to mammalian germ cells is relatively short (Russell *et al.*, 1984). Many compounds which cause somatic cell damage have been shown not to cause effects at gametogenesis (Holden, 1982). Furthermore, no compound has yet been demonstrated to give specific effects in germ cells without also showing effects in somatic cells.

It is thus doubtful that any germ cell chromosome assay should be adopted for routine screening. There is justification for restricting germ cell testing to special situations as the handling and analysis of such cells presents technical difficulties not encountered in the analysis of somatic cells.

When, however, germ cell data are needed for estimation of genetic risk, testing in the male can be performed in the mitotically proliferating spermatogonia, the stem cell population in particular being at long-term risk for producing heritable damage owing to the possible formation of clones. However, many chromosomal errors induced in pre-meiotic cells result in the death of the cell or prevent the passage of the cell through the meiotic divisions. Abnormalities such as translocations are much more likely to be efficiently transmitted to the F_1 progeny when induced in spermatids or spermatozoa (Albanese, 1987a).

In females, the oogonial divisions and prophase stages of meiosis up to the arrested diplotene stage occur during foetal ovarian development. Testing therefore requires the use of pregnant animals at appropriate stages of gestation. In the adult female, the most commonly tested stage is the arrested 'dictyate' oocyte because testing on more precisely selected times during the first or second meiotic divisions requires the use of natural or hormone-stimulated oocytes undergoing ovulation. The technical problems associated with the testing of female germ cells are therefore much greater than those involved in the male.

5.1.2.3 Analysis of heritable chromosome damage

The induction of heritable chromosome damage may be assayed by several methods, e.g. cytogenetic analysis of fertilised ova (reviewed in Albanese, 1987a); metaphase analysis of early embryos (Hansmann, 1973), or the heritable translocation test which involves the examination of diakinesis/metaphase I spermatocytes for multivalent association in adult male F_1 animals (Cattanach, 1982). The technique for the preparation of suitable material for analysis of early embryos is demanding (Rohrborn *et al.*, 1977) and consequently it is desirable to have some information available on the effects of fertility of a test compound before commencing an analysis of chromosome damage. Similar considerations apply to the heritable translocation test which has the additional complication of requiring considerable numbers of animals to achieve an adequate sample size for statistical evaluation.

Whereas a positive response in the heritable translocation test provides unequivocal evidence of induced heritable cytogenetic damage, the chromosome damage seen in fertilised ova or early embryos may be incompatible with post-natal survival (Brewen *et al.*, 1975) and therefore may not be indicative of potential to cause heritable chromosome damage.

5.1.2.4 Analysis of embryos following transplacental exposure

Chromosome damage can be assessed in embryonic cells following exposure of the pregnant animal (reviewed in Henderson, 1986). Cell preparations from whole embryos or from isolated embryonic tissues can be assessed by either metaphase analysis or induction of micronuclei. Micronuclei can be scored in polychromatic erythrocytes in the liver or peripheral blood of foetal mice (Cole *et al.*, 1981). Although it has been demonstrated that analysis of foetal tissue may be a more sensitive indicator of genotoxic effects for some compounds than the analysis of adult tissue (Henderson, 1986), the main reason for performing transplacental studies is to investigate specific embryonic risk factors.

5.2 EXPERIMENTAL DESIGN

5.2.1 General comments

Some aspects of study design will be common to all types of *in vivo* cytogenetic assay and these are discussed below. The number of animals per group will be discussed in Section 5.3.

5.2.2 Dose formulation

Water-soluble materials generally present few formulation problems and can be administered in distilled water or isotonic saline. Even if rapid or extensive hydrolysis is anticipated, this will probably be representative of any 'real-life' exposure conditions provided that all doses are freshly prepared and used promptly. The pH levels of aqueous solutions may require adjustment if they deviate markedly from physiological pH. Dissolution in isotonic saline seems to have little advantage over the use of water alone except when a solution of very low concentration is given by a parenteral route.

Selection of an appropriate vehicle for materials insoluble in water will depend on a number of factors.

- If the substance as supplied, or after milling, produces a stable and apparently homogeneous suspension in a water-based vehicle such as 0.5% w/v methyl cellulose or carboxymethylcellulose or equivalent, this should be used. Concentrations up to 2% w/v methyl cellulose or carboxymethylcellulose are commonly employed. Formulation using a high-shear mixer, with resuspension immediately before dosing, will frequently be acceptable for such materials.
- If immiscibility with water means that aqueous suspensions are not practicable, the test material may dissolve in corn (maize) oil. However, this may affect the rate at which the test material is absorbed after dosing.
- If neither of the above approaches is successful, other vehicles must be considered. The addition of a dispersant such as Tween 80 to aid suspension of non-wettable materials in aqueous formulations is permitted.

The use of any unusual or novel vehicle to the testing laboratory will necessitate the inclusion of an untreated control group, or a control group dosed with a vehicle used routinely, in addition to the vehicle controls. The use of organic solvents is not recommended. In particular dimethylsulphoxide should not be used as it markedly increases penetration through biological membranes and may invalidate normal patterns of absorption and distribution (as well as being locally irritant and toxic).

Exposure by inhalation poses technical problems which are outside the scope of this guideline, and an experienced inhalation toxicologist should be consulted if this route is to be used.

5.2.3 Preliminary toxicity assay

Before proceeding to the main study it is essential that the toxicity of the test substance be assessed in a preliminary toxicity assay. (Experience has shown that reliance on the results of an acute toxicity assay, where the substance is administered once with a 14-day observation period, may be insufficient to determine the maximum tolerated dose (MTD), in which case an independent toxicity test may be required.)

To satisfy regulatory guidelines (e.g. OECD Guidelines for the micronucleus and chromosome aberration assays, see Table 5.1) it is generally necessary to test chemicals at the maximum tolerated dose (MTD) or a dose which produces cytotoxic effects. Precise definitions of the MTD vary. It is used by the EEC as the highest dose level which produces signs of toxicity without having major effects on survival relative to the test in which it is used (Official Journal of the European Community Part B: Methods for the Determination of Toxicity 19.9.84). Preliminary toxicity assays to determine the maximum tolerated dose should be designed so as to encompass some of the following parameters; death, clinical signs of compound related toxicity (e.g. piloerection, hypothermia, ataxia and ptosis), change in body weight or cytotoxicity measured by the ratio of polychromatic to normochromatic cells or mitotic index. Toxicity is normally measured over a period of at least 72 hours, although it may be acceptable to chose an MTD based on clinical signs observed over a longer period of time.

5.2.4 Dose levels

This is a contentious area as the dose and/or.dose levels selected may depend on the purpose for which the data are intended. Table 5.1 summarises the variations in requirements by different regulatory authorities. A secondary consideration is whether the data are being generated for screening purposes or for more extensive safety evaluation (quantitative hazard assessment), to aid in risk assessment.

The question of an upper limit for the adequate evaluation of non-toxic chemicals was not addressed in the previous guidelines (Topham *et al.*, 1983). In the absence of toxic or cytotoxic effects we recommend that the highest dose chosen for *in vivo* cytogenetics assays should be 2 g/kg, consistent with the OECD and proposed EEC guidelines for acute oral toxicity testing. However, it should be noted that at the present time this dose level will not conform to some guidelines (e.g. OECD Guidelines for chromosome analysis in germ cells, Joint directive of the Japanese EPA, JMITI, JMHW, US EPA, see Table 5.1).

Studies using a single dose level are adequate for screening purposes. When data are required for more extensive safety evaluation, more than one dose level is advisable. The separation of the lower doses should be sufficient to determine whether a dose-response could be obtained. The precise levels will depend on the compound under evaluation and its intended use, but as a guide a two-fold separation factor should be considered.

5.2.5 Selection of species

At present, the mouse is most commonly employed for bone marrow micronucleus tests, and the rat for bone marrow metaphase analysis. However, both species can be used for either assay. The Chinese hamster has also been widely used for metaphase analysis studies because of its low chromosome number (2n = 22), but the absence of other toxicological or pharmacological data in this species and its cost usually argues against its use. A large number of compounds have been examined in the micronucleus assay in both rats and mice, and to date there are only two examples of a qualitative difference in response between these species (Albanese *et al.*, 1988). The selection of species should also be influenced by the availability of historical control data within each laboratory, and it may be relevant to note that the lower bodyweight of mice means that less test substance is needed for this species and that mice are cheaper and easier to handle than rats.

5.2.6 Route of administration

The investigator must consider two factors when choosing the route of administration: the need to maximise the possibility of absorption and penetration to target cells and to mimic human exposure (including accidental or environmental exposure). It must be borne in mind that these objectives will frequently lead to the selection of two different routes. Intraperitoneal (i.p.) and oral (gavage) routes have been most frequently used and the choice of route has been the subject of debate

Table 5.1 *Differences in guidelines of regulatory and other bodies*

Guideline	Number of animals per group	Sex	Number of dose levels	MD	Route	Sample times	Number of cells per animal	
							MN	CA
OECD (MN)[a]	10	m + f	1	MTD	NS	3	1000	
OECD (CA)[b]	10	m + f	1	MTD	i.p./oral	3		NS
OECD (germ)[c]	>5	m	1	5 g/kg	i.p./oral	3		100
OECD acute tox[d]	>5	m/f	3	2 g/kg†	—	—	—	—
EEC acute tox[e]	10	m + f	1	5 g/kg†	—	—	—	—
EEC[f]	10	m + f	1	5 g/kg	—	3	1000	50
JMHW[g]	5	m	3	2 g/kg	i.p./clin	3	1000	
JEPA/JMHW/MITI[h]	5	m	3	5 g/kg	i.p.	3	1000	
EPA genetox[i]	>6	m + f	3	50–80% LD_{50}	i.p./oral	3	2000	
EPA genetox[j]	>3	NS	3	‡	i.p.	3–4		50
EPA TOSCA[k]	10	m + f	1–3	MTD	i.p./oral	3	1000	—
ASTM[l]	10	m + f	3	5 g/kg	i.p.	3	*	50

* Depends on power and sensitivity required and on spontaneous control values.
† Under revision, a lower dose is likely to be recommended in the revised guidelines.
‡ No less than a factor of 2 less than a dose producing a significant level of toxicity.

MD, Maximum dose; NS, not specified; MN, micronucleus; CA, chromosome aberrations.

[a] OECD Guideline for Testing of Chemicals No. 474: Genetic Toxicology: Micronucleus test (1983).

[b] OECD Guideline for Testing of Chemicals No. 475: Genetic Toxicology: *in vivo* mammalian bone marrow cytogenetic test – chromosomal analysis (1984).

[c] OECD Guideline for Testing of Chemicals No. 483: Genetic Toxicology: Mammalian germ cell cytogenetics assay (1986).

[d] OECD Guideline for Testing of Chemicals No. 401: Acute Oral Toxicity (1987).

[e] Official Journal of the European Community Directive No. 84/449. 25 April 1984.

[f] Official Journal of the European Community No. L251 (19/9/1984). B. Methods for determination of toxicity. B12. Other effects, Mutagenicity, Micronucleus test.

[g] Ministry of Health and Welfare, Japan. Guidelines for Testing of Drugs for Toxicity. Pharmaceutical Affairs Bureau. Notice No. 118 (1984).

[h] Joint Directives of the Japanese Environmental Protection Agency, Japanese Ministry of Health and Welfare and Japanese Ministry of International Trade and Industry. 31 March 1987.

[i] Mavournin, K., Blakey, D., Cimino, M., Salamone, M. F. & Heddle, J. A. US Environmental Protection Agency Gene-Tox program, in preparation, 1989.

[j] Preston *et al.*, 1981.

[k] Toxic Substances Control Act. New and Revised Health Effects Test Guidelines. US Environmental Protection Agency (1983).

[l] ASTM American Society of Testing and Materials (MacGregor *et al.*, 1987; Preston *et al.*, 1987).

over the past few years. Ashby (1985) has questioned the continued use of the i.p. route, suggesting that its use undermined the advantages of using *in vivo* assays such as the micronucleus test 'as arbiters of whether *in vitro* genotoxins are likely to be absorbed, distributed and appropriately metabolized to produce genotoxic responses *in vivo*'. Shelby, on the other hand, argues that two routes (i.p. and the route relevant to human exposure) could be used sequentially during the evaluation of a novel compound (Shelby, 1986). In such situations it would be necessary to accept negative results from thorough second-stage studies as being better indicators of the risk associated with exposure to the chemicals, than the positive result obtained in a screening situation where the factors involved in the absorption, distribution and metabolism of the test compound are considered of lesser importance. A positive response in a test using the i.p. route followed by a negative response with oral dosing would require clarification using an alternative assay in another tissue (e.g. liver micronucleus test). The effect of using the oral or i.p. route has been compared using a number of compounds in a recent Japanese collaborative study (Hayashi *et al.*, 1989). They conclude that either route is acceptable and comparable in sensitivity when toxicity is accounted for.

In general, the oral route is recommended unless it is known that a chemical will be poorly absorbed using this route. In this case the options are two-fold: either the i.p. route can be used or an alternative assay can be used in which a more relevant target tissue can be examined. For risk assessment the use of the route most appropriate to human exposure should be used. The dermal or subcutaneous route should not be used unless it can be shown that the compound is taken up by this route.

Dose volumes should not normally exceed 10 ml/kg by the intravenous or intraperitoneal routes or 20 ml/kg given orally. In general, overnight starvation is not recommended.

5.2.7 Controls

All experiments should include a concurrent vehicle control group at each sampling time, against which treated groups will be compared. Use of a non-standard vehicle will necessitate the inclusion of untreated controls (see Section 5.2.3), but these are not normally required.

A positive control group should also be included. As the main purpose of this is to confirm the sensitivity of the assay (and to act as a check on scoring), it is not normally necessary for more than a single sample time to be used. The number of animals used in this group can be reduced. The dosage of a known clastogen used as the positive control should be set to avoid gross systemic toxicity and massive chromosome damage (so that some indication of reduced assay sensitivity can be observed). In the cases where a chemical structurally related to the test material is known to be clastogenic *in vivo*, it is advisable that this is included as a reference control.

5.2.8 Repeat experiments

As for other *in vivo* assays, tests for chromosome damage *in vivo* are performed only once. Technical failure or equivocal results in an otherwise valid study may, of course, make further testing necessary as may the need to examine an alternative route of exposure. If a second test is necessary, consideration should be given to any data which may indicate a need to re-design the study to optimise its sensitivity.

5.2.9 Animal husbandry and chemical safety precautions

Animal studies conducted in the UK must, of course, comply with all relevant Home Office regulations for animal husbandry and welfare, as well as for Personal and Project Licences. The Royal Society/UFAW Guidelines (1987) should be observed. Environmental conditions should be controlled and monitored.

Safety precautions necessary to protect personnel from exposure to known or suspect mutagens and/or carcinogens must be employed (e.g. Ehrenburg & Wachthmeister, 1977; Waters, 1980). These, together with waste disposal procedures to avoid environmental contamination, must take account of the special hazards associated with *in vivo* testing: mutagenic metabolites may be excreted by the test animals, and exposure by inhalation or by dietary administration (with associated risks of powder aerosols) requires special equipment and facilities.

5.3 THE PROCEDURE

5.3.1 Rodent micronucleus test

5.3.1.1 Introduction

As stated above, the micronucleus assay can be used as an initial screen to assess hazard (the potential to cause damage) or, by using a more elaborate protocol, be used for more extensive (quantitative) hazard assessment to aid in risk assessment of a chemical.

5.3.1.2 ***Species/sex***

Either the mouse or the rat can be used. It has been recently suggested that the rat offers substantial advantages over the mouse given that the majority of toxicological data on new chemicals and drugs are generated in the rat (Pascoe & Gatehouse, 1986; Albanese, 1987b). Relatively few rat studies had been carried out prior to 1987, mainly because of the presence of 'contaminating' mast cell granules within the marrow smears, but methods to avoid this are now available and discussed in Section 5.3.1.5.b.

Theoretically either species can be used routinely as most of the clastogens evaluated to date have been detectable using either the mouse micronucleus test or the rat bone marrow metaphase analysis assay (Thompson, 1986). However, recent findings indicate that species-specific effects can occur such that false negative results would be obtained if the 'wrong' species was selected (Albanese *et al.*, 1988). Since the data indicated that this may occur with only a very small number of chemicals, however, a recommendation that **both** species should be used routinely for the micronucleus test is considered inappropriate. Evidence of species-specific effects for a test chemical in other toxicological tests should be taken into account, and may influence the choice of species for the micronucleus assay. Further investigations into the possible existence of additional species differences are required as a matter of urgency.

Since the formulation of the previous UKEMS Guidelines on the conduct of the micronucleus assay (Topham *et al.*, 1983), there have been no published examples of qualitative sex differences in response to clastogens using the micronucleus assay, although quantitative differences have been shown to exist (Henry *et al.*, 1980; The Collaborative Study Group for the Micronucleus Test, 1986b). Consequently it should not be necessary to include both sexes in the study. This contrasts with the current guidelines for conduct of the micronucleus assay (EEC, OECD, EPA, see Table 5.1), all of which recommend the use of both sexes. It is especially noteworthy that the Japanese Ministry of Health and Welfare (JMHW, 1984; see Table 5.1) Guidelines specifically request that male animals only should be used. A reduction in the total number of animals obtained by using males alone would allow resources to be more gainfully employed on other aspects of the study design, for example in the accurate estimation of micronucleus incidence (see later), and possibly investigations into species specificity, if required.

Different strains of rats and mice appear to have different spontaneous micronucleated polychromatic erythrocyte (MPE) frequencies, but there are no data to suggest that qualitative differences exist in response to clastogens (The Collaborative Study Group for the Micronucleus Test, 1986a). Consequently there is no preferred strain although that selected should have a stable spontaneous incidence of micronuclei with a historical control mean of less than 4 per 1000 PCE (0.4%) as earlier advocated by Topham *et al.* (1983). The normal ratio of PCE to normocytes (NCE) within the marrow should be established. The use of specific hypersensitive strains for screening purposes has been advocated by some workers (Aeschbacher, 1986). While such strains may be useful for hazard identification their use is not recommended for risk assessment due to the difficulty in interpreting data obtained from hypersensitive animals.

Young adult animals should be used (e.g. 6–12 weeks for mice, 8–12 weeks for rats) and those of advanced age should be avoided because fat deposition in the marrow may reduce the clarity of stained preparations. The weight range for a particular experiment should be as narrow as possible.

5.3.1.3 Number of animals per group; number of erythrocytes analysed per animal

Topham *et al.* (1983) suggested that at least five animals should be used for each sampling point (with a minimum of four male and four female, if both sexes are justified). Recent investigations have indicated that more than five animals per dose group might be required on statistical grounds (Lovell *et al.*, 1989). Although regulatory guidelines generally require 10 animals per group (see Table 5.1) and 1000 PCE scored per animal, the analysis of 2000 PCE per animal would seem to allow a reduction in group size to approximately seven animals without a significant loss in test sensitivity (Lovell *et al.*, 1989).

It is important to emphasise that the **experimental unit is the animal and not the cell**. Consequently it is essential that sufficient PCE are scored to determine accurately the micronucleus incidence within the marrow of each individual animal. If too few cells (500) are examined, there is a danger that 'clusters' of MPE may be scored (Albanese & Middleton, 1987; Mirkova & Ashby, 1987).

The minimum number of PCE to score per animal also depends upon the historical control frequency for spontaneous micronucleus incidence as measured in each laboratory for the strain and species routinely used. The lower the spontaneous frequency, the greater the number of PCE per animal requiring analysis, to detect a specified level of increase in MPE. The existence of sampling differences between animals dictates that a **minimum** number of PCE should be analysed per animal. This is best achieved by employing a minimum number of animals per group

Table 5.2. *Overall power of an assay at different spontaneous MPE frequencies*

	Power of assay to detect a:			
	2-fold increase over control		3-fold increase over control	
Control MPE incidence/ 1000 PCE	7 animals/ group (%)	10 animals/ group (%)	7 animals/ group (%)	10 animals/ group (%)
1	46	57	84	94
2	70	82	98	100
3	84	94	100	100
4	92	98	100	100

(7–10) and scoring more than 1000 PCE per animal. The alternative of including more animals and scoring fewer cells per animal is less attractive on humane grounds and for the reasons previously discussed. The sensitivity of the assay is markedly affected by the PCE sample size and, in general, at least 1000 PCE should be scored per animal. However, 2000 PCE would give a more accurate estimate of micronucleus incidence, and should be used when the control incidence falls below 0.2%.

Classification/confirmation of equivocal results should be obtained by extending the slide assessment such that more than 2000 PCE are scored for MPE incidence before carrying out a repeat experiment.

Table 5.2 shows the overall power of an assay design in which 7 or 10 animals are used per group and 1000 PCE are scored per animal, in order to detect a doubling or trebling of the control micronucleus incidence. Power calculations are given assuming a Poisson distribution, for control micronucleus incidence of 1, 2, 3 and 4 micronuclei per 1000 PCE.

For positive-control treatments the group size could be reduced to a minimum of three animals, and only a single (near-optimal) sampling time needs to be used. The dose should be chosen such that the sensitivity of the test system can be monitored.

Assessment of the PCE:NCE (polychromatic erythrocyte: normochromatic erythrocyte) ratio can sometimes provide evidence that a sufficiently toxic dose has been administered, and that bone marrow cell exposure has taken place. Such effects result either from the inhibition of the division and maturation of nucleated erythropoietic cells, or the replenishment of the marrow with peripheral blood. In both cases a depression (real or apparent) of the proportion of PCE occurs. Once more the cell sample size is critical in obtaining an accurate estimate of this parameter. It has been suggested that the common practice of analysing 500 erythrocytes (PCEs + NCEs) for each animal is insufficient for this purpose (Mirkova & Ashby, 1987). It is recommended that if this end-point is regularly assessed, the sample size should be increased to at least 1000 total erythrocytes (PCE + NCE) to detect any compound related effects.

5.3.1.4 Dosing and sampling regimen

(a) *Choice of dosing regimen*

Dosing may be performed by using a single dose followed by multiple sampling times (single dose studies) or by using repeat dosing followed by sampling at a single time point (multi-dose studies). The single dose regimen has been most frequently used in the past although renewed interest is now being shown in multi-dose studies (MacGregor *et al.*, 1989; Tice *et al.*, 1989). The choice of treatment protocol must be made by the investigator on the basis of any available pharmacokinetic data on the test substance. The use of multiple dose regimens may be preferred if the test chemical is non-toxic or poses formulation problems. In this way a higher total dose can be given in two or more daily doses and this may allow the detection of weak clastogenic effects. Furthermore, the use of a multi-dose regimen would appear to be preferable for certain classes of potential clastogens, e.g. antimetabolites such as methotrexate (Yamamoto & Kikuchi, 1981; Hayashi *et al.*, 1984). Whilst some compounds are much more effective when a multi-dose regimen is used, e.g. dimethylbenzanthracene, azobenzene, the activity of some other compounds may be greatly reduced in a multi-dose study compared with a single dose study, e.g. dimethylhydrazine and hydroquinone. It can be concluded that neither dosing regimen will detect all clastogens detected by the other and therefore neither is recommended exclusively. Choice of procedure should be made on the basis of knowledge of the pharmacokinetic data, availability of toxicity data and experience of the laboratory with a particular protocol. It should be noted that multi-dose studies have the advantage of using fewer numbers of animals and hence a reduction in scoring time.

(b) *Single dose studies*

Since the life span of an immature (RNA positive) erythrocyte within the marrow is between 10 and 30 hours in both mouse and rat

(Salamone & Heddle, 1983) it is not necessary to sample the marrow of treated animals earlier than 19–24 hours after dosing (MacGregor *et al.*, 1987).

Nevertheless, most regulatory authorities require three sampling times, in the range 12–72 hours, and this is reiterated in the original UKEMS Guidelines (Topham *et al.*, 1983), theoretically to allow slowly absorbed or metabolised compounds to be detected or to overcome a compound-related delay in stem cell division. The need for later sampling times is based primarily upon published observations with the carcinogen 7,12-dimethylbenzanthracene (Bruce & Heddle, 1979; Salamone *et al.*, 1980; Salamone & Heddle, 1983). Recent results with this carcinogen have clearly shown that detectable and reproducible increases in micronucleus incidence occur at earlier sampling times (Proudlock & Allen, 1986; Ashby & Mirkova, 1987; Albanese, 1987b).

The need for a 72-hour sampling time is therefore considered unnecessary. It is believed that a reduction in sampling times to one around 24–30 hours, and another at approximately 48 hours after administration of test substance, would not reduce the sensitivity of the assay, nor impair its predictivity. Once more, available resources could be more usefully employed on the accurate analysis of micronucleus (MPE) incidence for each individual animal.

(c) Multiple dose studies

If multiple dose regimens (two or more doses) are used, the typical interval between doses is 24 hours. A sampling time of approximately 24 hours after the last dose is suitable for the detection of most chemicals currently known to induce micronuclei.

It may not always be possible to predict the dose levels required in a multi-dose study on the basis of single dose toxicity as enhanced toxicity of a chemical is frequently seen when administered repeatedly.

5.3.1.5 Slide preparation and analysis

The choice of method of slide preparation is one of individual preference. The most commonly used methods are listed below.

(a) Preparation of smears

The first and most widely practised method is to remove both femora; free them from extraneous muscle; insert the needle of a serum-containing syringe into the proximal part of each marrow canal; aspirate and then flush the shafts with serum such that a homogeneous cell suspension is obtained. The cells are then concentrated by gentle centrifugation (800–1000 rpm, 5 min), and a small drop of the viscous suspension is then spread on the surface of a clean microscope slide by pulling the material behind a cover glass held at 45° (Schmid, 1976).

An alternative method is to push the marrow directly onto the slide through a small opening in the iliac end of the femur by inserting a pin in the epiphysial end. The marrow is then mixed with serum to disperse the cells using the edge of a second clean microscope slide. Smears are made as before and air-dried (Salamone & Heddle, 1983).

A third technique has been described recently in which a fine sable paint brush, dipped in physiological saline, is inserted into the distal end of the femur and then drawn across a clean grease-free microscope slide. Four strokes are made across each slide such that the number of cells is gradually reduced (Albanese & Middleton, 1987).

(b) Staining procedures

For mouse bone marrow preparations. Romanowsky-type stains are routinely used, the most common of which is May-Grunwald/Giemsa (Schmid, 1976). Wrights' stain can also be used (Albanese & Middleton, 1987).

For rat bone marrow preparations, the presence of basophilic leukocyte (mast cell) granules, which stain in similar fashion to micronuclei, precludes the use of Romanowsky-type stains. However, the recent introduction of new staining procedures using either fluorescent stains (Hayashi *et al.*, 1983; MacGregor *et al.*, 1983), or haematoxylin and eosin (Pascoe & Gatehouse, 1986), which discriminate between true micronuclei and mast cell granules, allow the use of the rat in this assay. In addition, Romagna & Staniforth (1989) have published details of a method by which polychromatic and normochromatic erythrocytes can be concentrated and purified using a cellulose column, providing good quality preparations free of contaminating mast cell granules. All of these procedures allow discrimination between mast cell granules and true micronuclei.

(c) Slide analysis (scoring)

In this assay it is important to emphasise that the scored elements are the **micronucleated cells** and not the number of micronuclei.

To avoid observer bias all slides should be scored randomly under code by **one** investigator. It is essential that the definition of a micronucleus is clear in the analyst's mind prior to scoring, and that this is consistent within and between laboratories as far as practically possible. The majority of micronuclei are circular, within a PCE of normal morphology where it is coplanar with the cytoplasm. A minority of micronuclei might be oval or almond-shaped, and a few are ring-shaped. Doubtful bodies, through either morphology or staining characteristics, should not be scored; however, a note should be made of their presence as they may reflect other compound-related events taking place within the marrow.

In addition, the classification of erythrocytes as mature or immature can vary between laboratories. It is recommended that cells with a blue/red hue after Romanowsky staining (blue/grey after haematoxylin and eosin staining), should be classed as immature (PCE), whilst those unequivocally orange/red in colour should be classed as mature (NCE).

The smears should be analysed in a standard fashion by scanning the slide until areas containing cells with a suitable morphology and staining characteristics are encountered. It may be useful to note the vernier locations at the beginning and end of each scan across the slide, so that additional cells can be scored subsequently on the **same** slide if borderline or equivocal results are obtained initially, without fear of scoring the same cell population. Alternatively additional slides can be prepared for each animal so that the additional cells can be analysed on a different slide.

It is useful to assess the frequency of micronucleated normocytes for the purpose of quality control, since artefacts in any given slide will produce apparent increases in 'micronucleus' incidence in both normocytes and polychromatic erythrocytes. Typically the frequency of micronuclei in NECs should be 0–1/1000.

5.3.1.6 Data recording and presentation

For each animal the following data should be recorded: the number of PCEs, the number of NCEs and the number of micronucleated PCEs. In addition it may be useful to record the number of micronucleated NCEs. It is not necessary to record the vernier readings of micronucleated cells. The ratio of PCE:NCE should be calculated. Summary tables should be prepared showing the mean incidence of micronucleated PCEs and the PCE:NCE ratio at each dose level and sampling time for each treatment group. The mean incidence of micronucleated NCEs may also be tabulated. The results of the statistical analysis should also be presented.

5.3.1.7 Data evaluation

A positive response in the micronucleus test provides evidence that the test agent is capable of causing either chromosome breakage or chromosome loss in bone marrow erythroblasts of the rodent under the conditions of the study. The micronuclei arise from anaphase lag of chromosome fragments, bridged translocations or detached whole chromosomes.

The first step in data evaluation is the assessment of the negative controls, and then the positive controls for consistency with the expected values (i.e. should fall within the historical control range for that particular laboratory). Detailed recommendations on the type of statistical analysis to be carried out on data generated by this assay have been made previously (Lovell *et al.*, 1989), and the reader is referred to this publication.

If a statistically significant increase in the frequency of micronucleated PCEs is obtained at any dose or sampling time, a number of courses of action are open to the investigator.

1. If the response is weak or equivocal, additional PCE may be scored from each treated and control animal as described in Section 5.3.1.5(c) to confirm the result. If the response is real and not due to sampling error, statistical significance should increase with increased sample size.
2. A repeat study may be carried out using a range of dose levels and/or sampling times to 'optimise' the positive response, and possibly demonstrate a dose-response.
3. The accurate demonstration of a 'no-effect' dose level may be beneficial for certain classes of substances, to allow some form of 'risk-benefit' assessment to be made.

The definition of a negative response is more difficult. If the test material has been administered at the highest dose that can be formulated and administered, or at an upper limit of 2 g/kg, and a significant increase in micronucleated PCEs is not observed, an agent should be classified as negative under the conditions of the test.

5.3.2 Rodent bone marrow metaphase analysis

5.3.2.1 Species

Although the rat, mouse and Chinese hamster are all acceptable species, most bone marrow assays are carried out in the rat. This species has the advantage of being the most widely used in other toxicological investigations.

5.3.2.2 *Dose levels*

Most regulatory guidelines advocate the use of the maximum tolerated dose (MTD) (Section 5.2.4). This parameter by itself may prove inadequate and it is recommended that during the preliminary toxicity assay the MI_{50}, that dose causing a 50% reduction in the mitotic index, or some other parameter indicating cellular toxicity is established. For non-toxic substances an upper limit of 2 g/kg is recommended.

For screening purposes a single test group receiving the MTD may be satisfactory but in those instances where this assay will be used to provide data for quantitative hazard assessment a more elaborate protocol with at least three dose levels is recommended. If the MTD is the upper limit the lower dose levels should be separated by approximately two-fold intervals.

5.3.2.3 *Number of animals per group/number of cells scored*

Not all regulatory guidelines provide minimal criteria for conducting this assay as it has not always been considered a core test. The OECD Guideline (Table 5.1) suggests using five males and five females per group, whilst in the EPA genetox programme Preston *et al.* (1987) recommend that no fewer than three animals should be examined per experimental point. It is the opinion of this group that animals of one sex only may be used and that seven males per test group would be acceptable.

Concerning the number of cells scored, the OECD recommends a minimum of 50 cells per animal (500 per test group). If the number of animals is reduced (i.e. to seven as recommended above), it is nevertheless recommended that a minimum of 500 cells per test group is scored.

5.3.2.4 *Sampling interval*

The recommended sampling times in the OECD Guideline (Table 5.1) are 6, 24 and 48 hours after a single administration. The advantage of the 6-hour sample is that cells in their first mitosis will be available for analysis. It is unlikely that mutagens will be detected only at the 48-hour sample time and the need for this sample time is therefore questionable. It is suggested that it be used only when mitotic delay is observed. Effects on the cell cycle may be monitored during the preliminary toxicity assay when the need for a later sampling time can be evaluated.

5.3.2.5 Slide preparation and analysis

Two to four hours prior to sampling bone marrow animals should be dosed with a suitable mitotic arresting agent such as Colcemid or colchicine.

There are several excellent descriptions of methods that produce satisfactory spread and stained preparations of rodent bone marrow cells in metaphase (Dean, 1969; Adler, 1984). Preparations should be of adequate technical quality and all slides should be coded and randomised. The cytogenetic analysis of the whole study should be done blind by a competent trained observer. The vernier location of any cells with aberrations should be recorded. Cells with unusual or representative types of aberrations may be photographed. The record of observations made at the microscope constitute raw data for this type of study against which any observed or summarised data will be audited. Aberrations should be fully described in the raw data. The classification criteria of Scott *et al.* (Chapter 3) are recommended.

5.3.2.6 Data recording and presentation

The chromosome number should be recorded for all cells analysed and the vernier readings of all aberrant cells should be recorded. The following data should be recorded for each animal: number of cells scored, number of cells with each type of aberration, total number of cells with aberrations (including and excluding gaps), the frequency of aberrations per cell.

A summary table should also be prepared showing the mean results for each treatment group. It is recommended that data are presented as both frequency of aberrations per cell and percentage of cells with one or more aberrations **or** solely as the percentage of aberrant cells.

5.3.2.7 Data evaluation

A positive response in the rodent bone marrow assay provides evidence that the test substance is clastogenic under the conditions of the study.

In order to define that a positive response has been induced by the test material, it is necessary to satisfy pre-defined criteria. For example:

1. solvent and positive control responses should fall within an acceptable range for the laboratory;
2. a statistically significant difference between the treated and control groups should be demonstrated. Detailed recommendations on the type of statistical analysis to be carried out on data

generated by this assay have been made previously (Lovell *et al.*, 1989).

Tests that achieve these criteria are judged positive. Tests which do not fulfil the pre-determined criteria must be judged negative, or inconclusive when a repeat study may be needed.

5.3.3 Rodent germ cell assays

As discussed earlier (Section 5.1.2.2) rodent germ cell assays would not normally be conducted for routine screening purposes, but could be part of a package of data used in quantitative hazard assessment. Most of the comments in the preceding section (5.3.2) would apply equally to germ cell studies but the additional technical considerations pertinent to germ cell assays are provided.

Either the mouse or rat can be used although the mouse is frequently the preferred species. The air-drying method for the preparation of cells in the first and second meiotic metaphases (M I and M II) in the male (Evans *et al.*, 1964) does not work well with some other rodent species such as the rat and golden hamster. Although spermatogonial metaphases can be found on air-dried slides of the mouse prepared by this technique, their numbers can be boosted if the testicular tubules are dispersed in 0.25% trypsin prior to hypotonic treatment. For the treatment of spermatogonial stem cells and analysis at M I stage, sufficient time should elapse to allow the products of treatment to reach this stage of meiosis. In practice, at least 1 month between treatment and sample should be allowed to pass in the mouse. Timings for the treatment of spermatocytes at preleptotene and zygotene have been established by Brook & Chandley (1986) as 11 days 4 hours and 8 days 10 hours, respectively. Passage of cells from M I to M II requires 4 hours according to the same authors. These basic timings may, however, be subject to change in the case of compounds which alter the rate of spermatogenesis.

For detection of structural rearrangements induced in spermatogonia or during prophase, a search for multivalent formation can be made at M I. A more recent innovation, however, has been the introduction of a technique for analysis of the synaptonemal complex at pachytene in microspread preparations (Cawood & Breckon, 1983). Such pachytene preparations revealed nearly twice the number of multivalents found at M I in the Syrian hamster following an acute dose of X-rays to stem cell spermatogonia. However, examination of cells must be carried out by electron microscopy, and therefore may be unsuitable for many investigators.

For detection of errors of segregation at the first meiotic division in the male mouse, M II cells showing 19 (hypoploid) or 21 (hyperploid) chromosomes should be searched for (Brook & Chandley, 1986).

Few tests on the female mouse have been carried out, but technical detail of the procedures for inducing ovulation by hormones and for the chemical treatment of specific stages of meiosis in this species can be found, for example, in Hansmann & El-Nahass (1979), Brook (1982) and Brook & Chandley (1985).

5.4 SUMMARY AND CONCLUSIONS

The micronucleus test is recommended in preference to metaphase analysis in bone marrow cells for screening of chemicals for clastogenic potential *in vivo*.

Germ cell cytogenetic assays should be used only for risk assessment and should not form part of a screening battery.

The top dose level used in *in vivo* cytogenetic assays should be either the maximum tolerated dose, a dose which produces cytotoxic effects or 2 g/kg for a non-toxic substance.

Cytogenetic tests should be conducted initially in a single species (rat or mouse) with a single sex only. Seven to 10 animals per dose group should be used. Although 1000 PCEs per animal are normally scored in the micronucleus test, consideration should be given to scoring 2000 cells per animal to maximise the power of the assay. For metaphase analysis at least 500 cells per group should be analysed.

For the micronucleus test either a single dose study with two harvest times at 24 and 48 hours after treatment or a multiple dose study with two or three doses 24 hours apart, followed by a single harvest time 24 hours after the last dosing, is acceptable.

Metaphase analysis assays should be performed using a single treatment time with two harvest times in the period 6–24 hours after dosing.

EDITOR'S NOTE

At the time of preparation (1989) some of the recommendations made in this chapter (e.g. group size, number of harvest times) may not satisfy all published regulatory guidelines. However, it is recognised that these also may be revised. The reader is therefore advised to check current guidelines for specific requirements in conjunction with the recommendations made here.

5.5 REFERENCES

Adams, J.M., Gerondakis, S., Webb, E., Corcoran, L.M. & Cory, S. (1983). Cellular myc gene is altered by chromosome translocation to an immunoglobulin locus in murine plasmacytomas and is rearranged similarly in human Burkitt lymphomas. *Proceedings of the National Academy of Sciences (USA)*, **80**, 1982–6.

Adler, I-D. (1984). Cytogenetic tests in mammals. In *Mutagenicity Testing: a Practical Approach*, ed. S. Venitt and J.M. Parry. IRL Press, Oxford, pp. 275–306.

Aeschbacher, H.U. (1986). Rates of micronucleus induction in different mouse strains. *Mutation Research*, **164**, 109–15.

Albanese, R. (1987a). Mammalian male germ cell cytogenetics. *Mutagenesis*, **2**, 79–85.

Albanese, R. (1987b). The cytonucleus test in the rat: a combined metaphase and micronucleus assay. *Mutation Research*, **182**, 309–23.

Albanese, R. & Middleton, B.J. (1987). The assessment of micronucleated polychromatic erythrocytes in rat bone marrow. Technical and statistical considerations. *Mutation Research*, **182**, 323–33.

Albanese, R., Mirkova, E., Gatehouse, D. & Ashby, J. (1988). Species-specific response to the rodent carcinogens 1,2-dimethylhydrazine and 1,2-dibromochloropropane in rodent bone marrow micronucleus assays. *Mutagenesis*, **3**, 35–8.

Ashby, J. (1985). Is there a continuing role for the i.p. injection route of exposure in short term rodent genotoxicity assays? *Mutation Research*, **156**, 239–43.

Ashby, J. & Mirkova, E. (1987). A re-evaluation of the need for multiple sampling times in the mouse bone marrow micronculeus assay. Results for dimethylbenzanthracene. *Environmental Mutagenesis*, **10**, 297–305.

Ashby, J., Shelby, M.D. & de Serres, F.J. (1988). The IPCS *in vivo* collaborative study: overview and conclusions. In *Evaluation of Short-term Tests for Carcinogens. Report of the International Programme on Chemical Safety's Collaborative Study on* in vivo *Assays*, ed. J. Ashby, F.J. De Serres, M.D. Shelby, B.H. Margolin, M. Ishidate and G.C. Becking. pp. 1.6–1.28. Cambridge University Press, Cambridge.

Braithwaite, I. & Ashby, J. (1988). A non-invasive micronucleus assay in rat liver. *Mutation Research*, **203**, 23–32.

Brewen, J.G., Payne, H.S., Jones, K.P. & Preston, R.J. (1975). Studies on chemically induced dominant lethality. The cytogenetic basis of MMS induced dominant lethality in post-meiotic germ cells. *Mutation Research*, **33**, 239–50.

Brook, J.D. (1982). The effect of 4CMB on germ cells of the mouse. *Mutation Research*, **100**, 305–8.

Brook, J.D. & Chandley, A.C. (1985). Testing of 3 chemical compounds for aneuploidy induction in the female mouse. *Mutation Research*, **157**, 215–20.

Brook, J.D. & Chandley, A.C. (1986). Testing for the chemical induction of aneuploidy in the male mouse. *Mutation Research*, **164**, 117–25.

Bruce, W.R. & Heddle, J. (1979). The mutagenic activity of 61 agents as determined by the micronucleus, *Salmonella* and sperm abnormality assays. *Canadian Journal of Genetics and Cytology*, **21**, 319–34.

Cattanach, B.M. (1982). The heritable translocation test in mice. In *Cytogenetic Assays of Environmental Mutagens*, ed. T.S. Hsu. Allanheld, Osmun and Co. Totowa, New Jersey, pp. 289–323.

Cawood, A.D. & Breckon, G. (1983). Synaptonemal complexes as indicators of induced structural change in chromosomes after irradiation of spermatogonia. *Mutation Research*, **122**, 149–54.

Cole, R.J., Taylor, N., Cole, J. & Arlett, C.F. (1981). Short term tests for transplacentally active carcinogens. I. Micronucleus formation in foetal and maternal mouse erythroblasts. *Mutation Research*, **80**, 141–57.

Dean, B.J. (1969). Chemical-induced chromosome damage. *Laboratory Animals*, **3**, 151–74.

Ehrenburg, L. & Wachthmeister, C.A. (1977). Safety precautions in work with mutagenic and carcinogenic chemicals. In *Handbook of Mutagenicity Test Procedures*, ed. B.J. Kilbey, M. Legator, W. Nichols, and C. Ramel. Elsevier, Amsterdam–New York–Oxford, pp. 401–10.

Evans, E.P., Breckon, G. & Ford, C.E. (1964). An air-drying method for meiotic preparations from mammalian testes. *Cytogenetics and Cell Genetics*, **3**, 289–94.

Gilbert, F. (1983). Chromosomes, genes and cancer: A classification of chromosome abnormalities in cancer. *Journal of the National Cancer Institute*, **71**, 1107–14.

Hansmann, I. (1973). Induced chromosomal aberrations in pronuclei, 2-cell stages and morulae of mice. *Mutation Research*, **20**, 353–67.

Hansmann, I. & El-Nahass, E. (1979). Incidence of non-disjunction in mouse oocytes. *Cytogenetics and Cell Genetics*, **24**, 115–21.

Hayashi, M., Sofuni, T. & Ishidate, M. (1983). An application of acridine orange fluorescent staining to the micronucleus test. *Mutation Research*, **120**, 241–7.

Hayashi, M., Sofuni, T. & Ishidate, M. (1984). A pilot experiment for the micronucleus test. *Mutation Research*, **141**, 165–9.

Hayashi, M., Sutou, S., Shimada, H., Sato, S., Sasaki, Y.F. & Wakata, A. (1989). Difference between intraperitoneal and oral gavage application in the micronucleus test. The 3rd collaborative study by CSGMT/JEMS.MMS. *Mutation Research*, **223**, 329–44.

Henderson, L. (1986). Transplacental genotoxic agents: cytogenetic methods for their detection. In *Chemical Mutagens. Principals and Methods for their Detection, Vol. 10*, ed. F.J. de Serres. Plenum Press, New York, pp. 327–55.

Henry, M., Lupo, S. & Szabo, K.T. (1980). Sex difference in sensitivity to the cytogenetic effects of ethyl methane sulphonate in mice demonstrated by the micronucleus test. *Mutation Research*, **69**, 385–7.

Holden, H.E. (1982). Comparison of somatic and germ cell models for cytogenetic screening. *Journal of Applied Toxicology*, **2**, 196–200.

JMHW (1984). *Guidelines for Testing of Drugs for Toxicity*. Pharmaceutical Affairs Bureau, Notice No. 118. Ministry of Health and Welfare, Japan.

Kliesch, U., Danford, N. & Adler, I-D. (1981). Micronucleus test and bone marrow chromosome analysis. A comparison of two methods *in vivo* for evaluating chemically induced chromosome alterations. *Mutation Research*, **80**, 321–32.

Lovell, D.P., Anderson, D., Albanese, R., Amphlett, G.E., Clare, G., Ferguson, R., Richold, M., Papworth, D.G. & Savage, J.R.K. (1989). Statistical analysis of *in vivo* cytogenetics assays. In *UKEMS Sub-committee on Guidelines for Mutagenicity Testing. Report. Part III. Statistical Evaluation of Mutagenicity Test Data*, ed. D.J. Kirkland. Cambridge University Press, Cambridge, pp. 184–232.

MacGregor, J.T., Heddle, J.A., Hite, M., Margolin, B., Ramel, C., Salamone, M.R., Tice, R.R. & Wild, D. (1987). Guidelines for the conduct of micronucleus assays in mammalian bone marrow erythrocytes. *Mutation Research*, **189**, 103–12.

MacGregor, J.T., Wehr, C.M. & Gould, D.H. (1980). Clastogen-induced micronuclei in peripheral blood erythrocytes: The basis of an improved micronucleus test. *Environmental Mutgenesis*, **2**, 509–14.

MacGregor, J.T., Wehr, C.M., Henika, P.R. & Shelby, M.D. (1989). The *in vivo* micronucleus test. Measurement at steady state increases assay efficiency and permits integration with toxicity studies. *Fundamentals of Applied Toxicology*. (in press).

MacGregor, J.T., Wehr, C.M. & Langlas, R.G. (1983). A simple fluorescent staining procedure for micronuclei and RNA in erythrocytes using hoechst 33258 and pyronin Y. *Mutation Research*, **120**, 269–75.

Mirkova, E. & Ashby, J. (1987). Relative distribution of mature erythrocytes, polychromatic erythrocytes (PE) and micronucleated PE in mouse bone marrow smears: Control observations. *Mutation Research*, **182**, 203–11.

Newton, M.F. & Lilly, L.J. (1986). Tissue specific clastogenic effects of chromium and selenium salts *in vivo*. *Mutation Research*, **169**, 61–9.

Pascoe, S. & Gatehouse, D. (1986). The use of a simple haematoxylin and eosin staining procedure to demonstrate micronuclei within rodent bone marrow. *Mutation Research*, **164**, 237–43.

Preston, R.J., Au, W., Bender, M.A., Brewen, J.G., Carrano, A.V., Heddle, J.A., McFee, A.F., Wolff, S. & Wassom, J.S. (1981). Mammalian *in vivo* and *in vitro* cytogenetic assays: A report of the U.S. EPA's Gene-Tox Program. *Mutation Research*, **87**, 143–88.

Preston, R.J., Dean, B.J., Galloway, S., Holden, H., McFee, A.F. & Shelby, M.D. (1987). Mammalian *in vivo* cytogenetics assay: Analysis of chromosome aberrations in bone marrow cells. *Mutation Research*, **189**, 157–66.

Proudlock, R. & Allen, J. (1986). Micronuclei and other nuclear anomalies induced in various organs by DEN and DMBA. *Mutation Research*, **174**, 141–3.

Rohrborn, G., Hansmann I., & Buckel, U., (1977). Cytogenetic analysis of pre and post-ovulatory oocytes and pre-implantation embryos in mutagenesis of mammals. In *Handbook of Mutagenicity Test Procedures*, ed. B.J. Kilbey, M.S. Legator, W. Nichols and C. Ramel. Elsevier, Amsterdam, pp. 301–10.

Romagna, F. & Staniforth, C.D. (1989). The automated bone marrow micronucleus test. *Mutation Research*, **213**, 91–104.

Royal Society/UFAW (1987). *Guidelines on the Care of Laboratory Animals and their Use for Scientific Purposes*. Royal Society and the Universities' Federation for Animal Welfare, London.

Russell, L.B., Aaron, C.S., de Serres, F., Generoso, W.M., Kannan, K.L., Shelby, M., Springer, J. & Voytek, P. (1984). Evaluation of mutagenicity assays for purposes of genetic risk assessment. *Mutation Research*, **134**, 143–57.

Salamone, M.F., & Heddle, J. (1983). The bone marrow micronucleus assay: rationale for a revised protocol. In *Chemical Mutagens. Principles and Methods for their Detection, Vol. 8*, ed. F.J. de Serres. Plenum Press, New York, pp. 111–49.

Salamone, M.F., Heddle, J., Stuart, E. & Katz, M. (1980). Towards an improved micronucleus test. Studies on 3 model agents, mitomycin C, cyclophosphamide, DMBA. *Mutation Research*, **24**, 347–56.

Schmid, W. (1976). The micronucleus test for cytogenetic analysis. In *Chemical Mutagens. Principles and Methods for their Detection, Vol. 4*, ed. A. Hollaender. Plenum Press, New York, pp. 31–53.

Shelby, M.D. (1986). A case for the continued use of the intraperitoneal route of exposure. *Mutation Research*, **170**, 169–71.

Shtivelman, E., Lifshitz, B., Gale, R.P. & Canaani, E. (1985). Fused transcript of ab_l and *bcr* genes in chronic myelogenous leukaemia. *Nature*, **315**, 550–4.

Tates, A.D., Neuteboom, I., Hofker, M. & Den Engelese, L. (1980). A micronucleus technique for detecting clastogenic effects of mutagens/carcinogens (DEN, DMN) in hepatocytes of rat liver *in vivo*. *Mutation Research*, **74**, 11–20.

The Collaborative Study Group for the Micronucleus Test (1986a). Sex differences in the micronucleus test. *Mutation Research*, **172**, 151–63.

The Collaborative Study Group for the Micronucleus Test (1986b). Strain differences in the micronucleus test. *Mutation Research*, **204**, 307–16.

Thompson, E. (1986). Comparison of *in vivo* and *in vitro* cytogenetic assay results. *Environmental Mutagenesis*, **8**, 753–67.

Tice, R., Erexson, G., Hilliard, C.J. & Shelby, M.D. (1989). Evaluation of treatment protocol and sample time on micronuclei frequencies in mouse bone marrow and peripheral blood. *Mutagenesis* (in press).

Topham, J., Albanese, R., Bootman, J., Scott, D. & Tweats, D. (1983). *In vivo* cytogenetic assays. In *UKEMS Sub-committee on Guidelines for Mutgenicity Testing. Report. Part I. Basic Test Battery*, ed. B. Dean. United Kingdom Environmental Mutagen Society, Swansea, pp. 119–41.

Tsuchimoto, T. & Matter, B.E. (1979). *In vivo* cytogenetic screening methods for mutagens with special reference to the micronucleus test. *Archives of Toxicology*, **42**, 239–48.

Waters, D.B. (ed.) (1980). *Safe Handling of Carcinogens, Mutagens, Teratogens and Highly Toxic Substances, 1*. Ann Arbor Science, USA.

Yamamoto, K. & Kikuchi, Y. (1981). Studies on micronuclei time response and on the effects of multiple treatments of mutagens on induction of micronuclei. *Mutation Research*, **90**, 163–73.

Yunis, J.J. (1983). The chromosomal basis of human neoplasia. *Science*, **221**, 227–36.

INDEX